INTERPRETING ENGINEERING DRAWINGS

C. JENSEN
TECHNICAL DIRECTOR
IN THE PROVINCE OF ONTARIO

R. HINES
ELECTRO MECH. SPECIALIST
C. G. E.

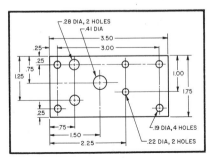

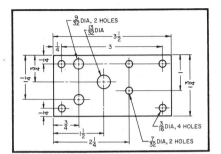

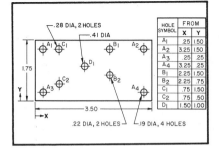

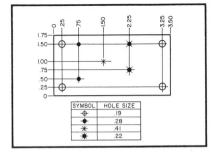

DELMAR PUBLISHERS • ALBANY, NEW YORK 12205

PREFACE

New technological developments in industry have created a need for well-prepared instructional material for persons entering or engaged in technical positions in industry. *Interpreting Engineering Drawings* provides instructional material in the correct sequence for students, apprentices, journeymen, and others who need to know how to read and interpret engineering drawings.

New drafting conventions, which have been brought about by the advances in new production techniques, and simplified drafting standards are used throughout the text. The use of a second color has been introduced to make the written material and the illustrations more understandable to the reader by providing improved readability, contrast, and emphasis.

The written material and illustrations which precede the drawing assignments contain related technical information and principles of drafting necessary to interpret the new material on each drawing. In some instances shop practices are defined and explained to clarify the meanings of operational notes which appear on shop drawings.

Selected tables and charts taken from handbooks and manufacturers' catalogs are located in the back of the book. These, in most part, comprise the reference material needed to solve the problems. Further information can be obtained by writing to the United States of America Standards Institute; 10 East 40th Street; New York, N.Y. 10016 and the Canadian Standards Association; 235 Montreal Road; Ottawa 7, Canada.

To distinguish between the illustrations used in the explanatory material and the assignments, the illustrations are given a unit and a figure number and the assignments are given a drawing number preceded by the letter "A". *Figure 1-1,* for example, is the first illustration in the book, and *Drawing A-1* is the first assignment.

The authors would like to express their appreciation to McGraw-Hill Company of Canada, Ltd., for giving permission to include pertinent illustrations published in *Engineering Drawing and Design* by Cecil H. Jensen.

Cecil H. Jensen and Raymond D. Hines

about the
AUTHORS

The choice of authors to revise and modernize the *Blueprint Reading Series for the Mechanical Trades* was part of a well-conceived plan to combine the latest trends of industrial design with the progressive teaching techniques in technical education.

Mr. Raymond D. Hines contributes to this book thirty years' experience in the field of electromechanical design. He is presently employed as a design specialist by the General Electric Company, Guelph, Ontario, where he was recently responsible for the internal design layout of the largest high-voltage power transformer built in North America. During this time he has been intensely interested in the advancement of the drafting design profession through the latest methods of drawing presentations, especially in the field of simplified drafting.

Mr. Cecil H. Jensen, technical director in the educational system of the Province of Ontario, Canada, has had over eighteen years' teaching experience in mechanical drafting. He is the successful author of many technical books including *Engineering Drawing and Design, Drafting Fundamentals,* and *Home Planning and Design.* Before entering the teaching profession, Mr. Jensen gained several years of design experience in industry. He has also been responsible for the supervision of the teaching of technical courses for the General Motors apprentices in Canada.

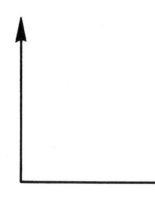

CONTENTS

Note: There is graph paper at the back of the book for sketching problems.

The author and editorial staff at Delmar Publishers are interested in continually improving the quality of this instructional material. The reader is invited to submit constructive criticism and questions. Responses will be reviewed jointly by the author and source editor. Send comments to:

Editor-in-Chief
Box 5087
Albany, New York 12205

UNIT 1

BASES FOR INTERPRETING DRAWINGS

Engineering or technical drawings are made for the purpose of furnishing a description of the shape and size of an object and other information necessary for its construction in such a form that it can be readily recognized by anyone familiar with engineering drawings. A picture or photograph of an object shows the object as it appears to the observer; it does not show the exact size, shape, and location of the various parts of the object. For this reason, a number of views are necessary, each showing that part of the object as it may be seen by looking directly at each one of the surfaces, figure 1-1, and then arranging these views in a systematic manner, projected one from the other. This type of projection is called orthographic projection. The understanding and ability to visualize an object from these views is essential in interpreting engineering drawings.

The principles of orthographic projection can be applied in four different *angles,* or systems: first, second, third, and fourth angle projection. Third-angle orthographic projection is used in North America and in many European countries; first angle is used primarily in the British Isles.

THIRD–ANGLE PROJECTION [1]

The third-angle system of projection is used almost exclusively on mechanical engineering drawings. This is because it involves much less drafting time than other methods and permits each facet of the object to be drawn without distortion of form, and to be drawn in true scale along all dimensions.

Usually, three views are sufficient to explain the shape of the object. The commonest views are the front, top, and right side. In this system the object may be assumed to be enclosed in a glass box, as in figure 1-2,

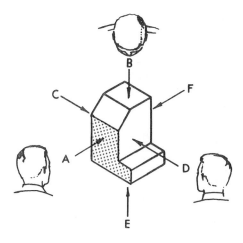

Fig. 1-1   Pictorial View of Object.
(Courtesy CSA B 78.1 – 1967)

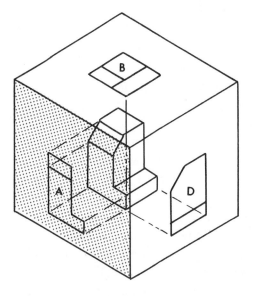

Fig. 1-2   Object Enclosed in Glass Box.
(Courtesy CSA B 78.1 – 1967)

1

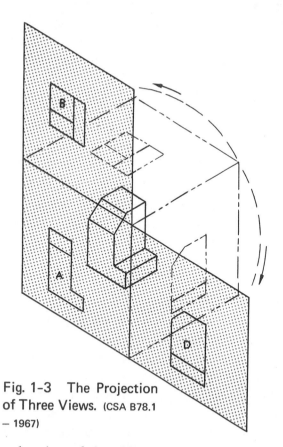

Fig. 1-3   The Projection of Three Views. (CSA B78.1 — 1967)

VISIBLE OBJECT LINES

The object line, a thick solid line, is used to indicate the visible edges and corners of an object. Object lines should stand out clearly in contrast to other lines so that the general shape of the object is apparent to the eye.

LETTERING ON DRAWINGS

The most important requirements for lettering are legibility, reproducibility, and ease of execution. These requirements are best met by the style of lettering known as standard uppercase Gothic, as shown in figure 1-6. For these reasons it is used on mechanical engineering drawings. Vertical lettering is preferred, but sloping style may be used, though never on the same drawing. Suitable lettering size for notes and dimensions is .12 (or 1/8) inch. Larger characters are used for drawing titles and numbers and where it may be necessary to bring some part of the drawing to the attention of the reader.

SKETCHING

Sketching is a necessary part in the course of interpreting technical drawings since the skilled craftsman in the shop is frequently called upon to sketch and explain his thoughts

and a view of the objects drawn on each side of the box represent that which is seen when looking perpendicularly at each face of the box. If the box were unfolded as if hinged around the front face, the desired orthographic projection would result, as shown in figures 1-3 and 1-4. These views are identified by names as shown.

The front, rear, and side views are sometimes called elevations (front elevation). The top view may be termed the plan. The bottom view is the view looking up at the object. Rear views may be shown at the extreme right.

Figure 1-5 shows a simple object drawn both in orthographic projection and pictorial form. The drawing has been made showing each side to represent the exact shape and size of the object and the relationship of the three views to one another. This principle of projection is used throughout all mechanical drawing. The isometric drawing shows the relationship of the front, top, and side surfaces.

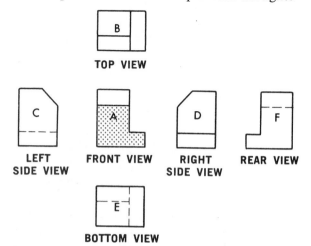

Fig. 1-4   Third-Angle Orthographic Projection: The Six Principal Views. (Courtesy CSA B 78.1 – 1967).

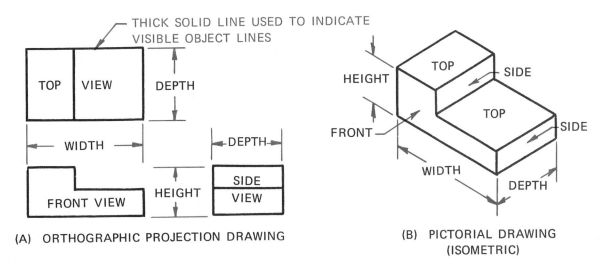

THICK SOLID LINE USED TO INDICATE VISIBLE OBJECT LINES

(A) ORTHOGRAPHIC PROJECTION DRAWING

(B) PICTORIAL DRAWING (ISOMETRIC)

Fig. 1-5   A Simple Object Shown in Orthographic and Pictorial Form.

ABCDEFGHIJKLMNOP QRSTUVWXYZ 1234567890

VERTICAL

ABCDEFGHIJKLMNOP QRSTUVWXYZ 1234567890

SLOPED

Fig. 1-6   Lettering for Drawings.

to other people. Sketching also helps to develop a good sense of proportion and accuracy of observation. The most common types of sketching paper are shown in figure 1-7. Each square on the paper may represent 1/10, 1/8, 1/4, 1/2, 1 inch or 1 foot of actual object length. Figure 1-7 illustrates the use of graph paper for both pictorial and orthographic projection.

REFERENCES AND SOURCE MATERIALS

1. Canadian Standards Association, *Bulletin* 78.1 (1967).

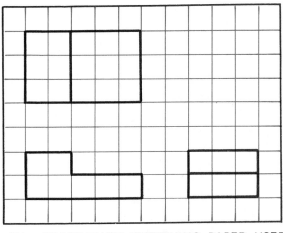

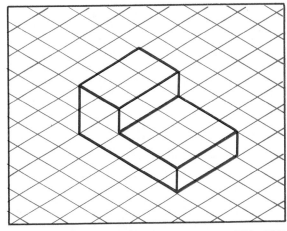

(A) COORDINATE SKETCHING PAPER USED FOR SKETCHING ORTHOGRAPHIC PROJECTION

(B) ISOMETRIC SKETCHING PAPER USED FOR SKETCHING PICTORIAL DRAWINGS

Fig. 1-7   Sketching Paper.

3

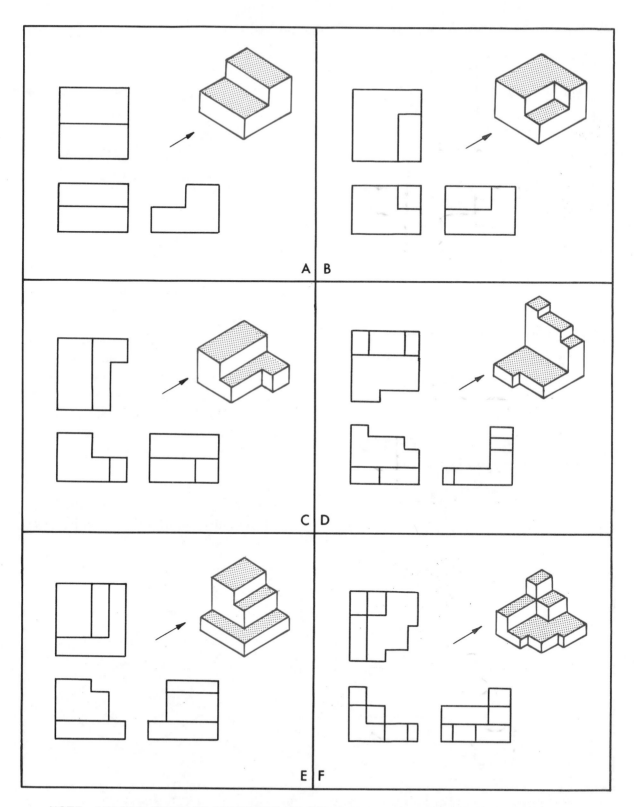

NOTE: ARROWS INDICATE DIRECTION OF SIGHT WHEN LOOKING AT THE FRONT VIEW.

Fig. 1-8   Illustrations of Simple Objects Drawn in Orthographic Projection.

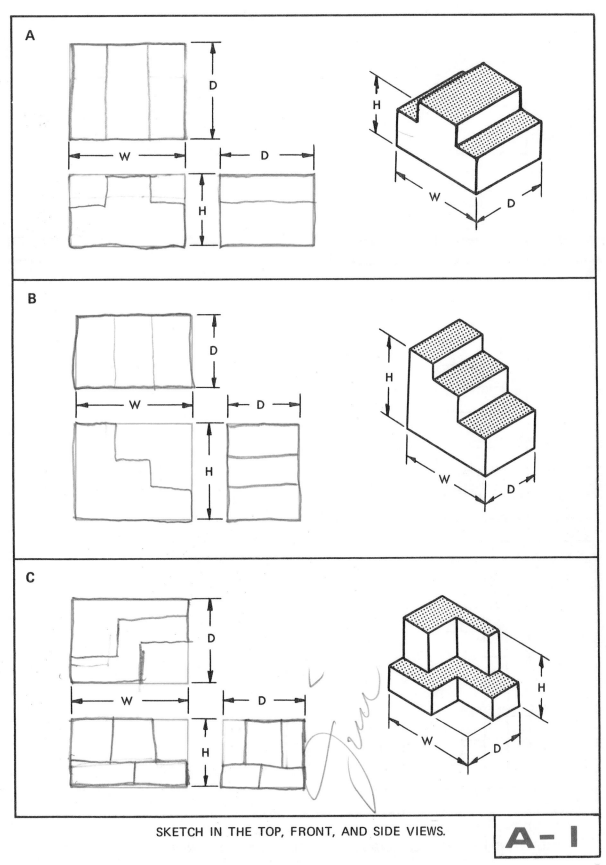

A

B

C

SKETCH IN THE TOP, FRONT, AND SIDE VIEWS.

A-1

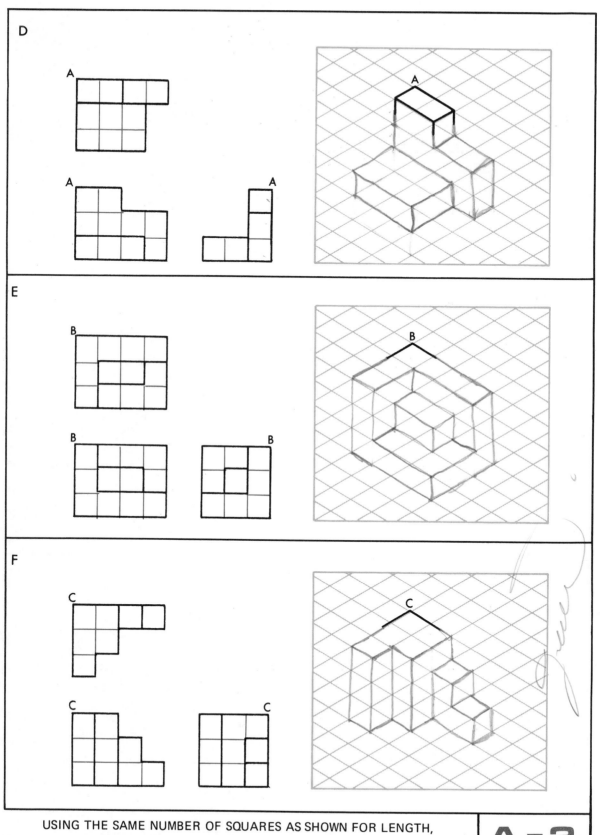

USING THE SAME NUMBER OF SQUARES AS SHOWN FOR LENGTH, WIDTH AND HEIGHT SIZES, COMPLETE THE PICTORIAL DRAWINGS.

A-2

UNIT 2

WORKING DRAWINGS

A working drawing, sometimes referred to as a detail drawing, is the drawing which is sent from the drafting office to the shop. It contains the information necessary to construct or assemble the part. The original drawing is kept in the office. The craftsmen in the shop work from copies called whiteprints or blueprints of the original drawing. Although the blueprint is still used in certain areas and preferred by some facets of industry, it is rapidly being replaced by the whiteprint. The original drawing must, therefore, be on a type of translucent paper from which as many reproductions as desired can be made.

The working drawing must supply the complete information for the construction of the part. This information may be classified under three headings:

- Shape description (the number and type of view selected to completely show or describe the shape of the part),

- Size description (the dimensions which show the size and location of the shape of features).

- Specifications (general notes, material, heat treatment, finish, number required).

This data may be found on the drawing or in the title strip or block.

DIMENSIONS [1]

The size of an object is shown by placing measurements, called dimensions, on the drawing but preferably off the views.

Dimension Lines

Dimension lines are used to denote the extent of the dimension and should be drawn parallel to the dimension to which they apply. They terminate in arrowheads which just touch the extension line and are broken to insert the dimension. Unbroken lines are sometimes used as a simplified drafting practice. Where space does not permit the placement of the dimension line and dimension between the extension lines, the dimension line may be placed outside the extension line. The dimension can also be placed outside the extension line if the space between the extension lines is limited, figure 2-1.

Extension Lines

Extension lines are used to denote the points or surfaces between which a dimension applies. They extend from object lines and are drawn perpendicular to the dimension lines, figure 2-1. A small gap is left between the extension line and the outline to which it refers.

Leaders

Leaders are used to direct dimensions or notes to the surface or points to which they apply. A leader consists of a short horizontal bar adjacent to the note or dimension and an inclined portion which terminates with an arrowhead touching the line or point to which it applies, or which terminates with a dot when it refers to a surface within the outline of a part.

Dimensioning Units

It is recommended that all linear dimensions other than metric be expressed in inches and preferably in the decimal system. However, the fractional inch system and the use of feet and inches are used to a limited extent.

Inch marks are not shown in the decimal or fractional inch systems. In the decimal inch system, parts are designed in basic decimal increments, preferably .02 inch, and are ex-

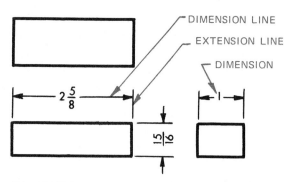

(A) ALIGNED SYSTEM OF DIMENSIONING USING FRACTIONAL UNITS OF MEASUREMENTS

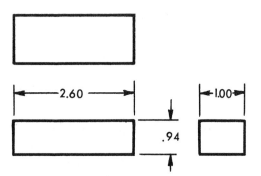

(B) UNIDIRECTIONAL SYSTEM OF DIMENSIONING USING DECIMAL SYSTEM OF DIMENSIONING

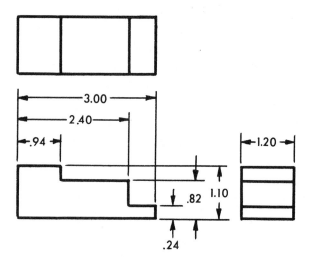

(C) PLACEMENT OF DIMENSIONS

Fig. 2-1 Basic Dimensioning Elements.

pressed as two-place decimal numbers unless greater accuracy is required. The decimal point should be bold and clear. The dimensions may be placed on the drawing so that all dimensions read from the bottom (unidirectional system), or the dimensions may be placed perpendicular to the dimension line (aligned system), figure 2-1. The unidirectional system of dimensioning is now preferred.

CHOICE OF DIMENSIONS

The choice of the most suitable dimensions and dimensioning methods will depend, to some extent, on whether the drawings are intended for unit production or mass production.

Unit production refers to cases in which each part is to be made separately, using general-purpose tools and machines. Details on custom-built machines, jigs, fixtures, and gauges required for the manufacture of production parts are made in this way. Frequently, only one of each part is required.

Mass production refers to parts produced in quantity, where special tooling is usually provided. Most part drawings for manufactured products are considered to be for mass-produced parts.

Functional dimensioning should be expressed directly on the drawing, especially for mass-produced parts. This will result in the selection of datum features on the basis of function and assembly. For unit-produced parts, it is generally preferable to select datum features on the basis of manufacture and machining.

Nonfunctional dimensions should be selected on the basis of facilitating production and inspection rather than to facilitate tool and gauge design.

INFORMATION SHOWN ON ASSIGNMENT DRAWINGS

Letters are used on the drawing assignments to follow so that questions may be asked about lines and surfaces without involving a great number of descriptive items. They are learning aids and, as the course progresses, are omitted from the more advanced problems. To simplify the drawings, the actual working drawing is drawn in black. The information which is used in the developing of interpreting technical drawings is shown in red and would not appear on working drawings found in industry.

REFERENCES AND SOURCE MATERIALS

1. Canadian Standard Association, *Bulletin* 78.2 (1967).

QUESTIONS

1. What is the name of the object?
2. What is the drawing number?
3. How many pieces are to be made?
4. Of what material is the part made?
5. What is the overall length?
6. What is the overall width?
7. What is the overall height or thickness?
8. What line in the side view represents surface (F) in the top view?
9. What line in the side view represents surface (E) in the top view?
10. What line in the side view represents surface (G) in the top view?
11. What line in the side view represents surface (L) of the front view?
12. What is the vertical height in the side view from the surface represented by line (P) to that represented by line (Q) ?
13. What is the height of the step in the side view from the bottom of the bar to the line representing surface (E) ?

14. What two letters in the top view represent distance (V) in the side view?
15. What two letters in the top view represent distance (W) in the side view?
16. What line in the side view represents surface (M) in the front view?
17. What is the height of line (N) ?
18. What line in the front view represents the surface (R) in the side view?
19. What line in the top view represents surface (L) ?
20. What line in the front view represents surface (F) ?
21. What line in the front view represents surface (E) ?
22. What line in the top view represents surface (M) ?
23. What type of line is (T) ?
24. What type of line is (Y) ?
25. What type of dimensions are used on this drawing?
26. Calculate dimensions **B, C, D, V,** and **W.**

Answers (handwritten):

1. counter Clamp
2. A-3
3. 2
4. M.S
5. 2.50
6. 1.12
7. .50
8. Q
9. P
10. S
11. T
12. .18
13. .32
14. C - D
15. B - C
16. N
17. .32
18. A
19. H
20. S
21. K
22. X
23. object li[ne]
24. Dimension
25. decimal
26. B .38
    C .50
    D .24
    V .74
    W .88

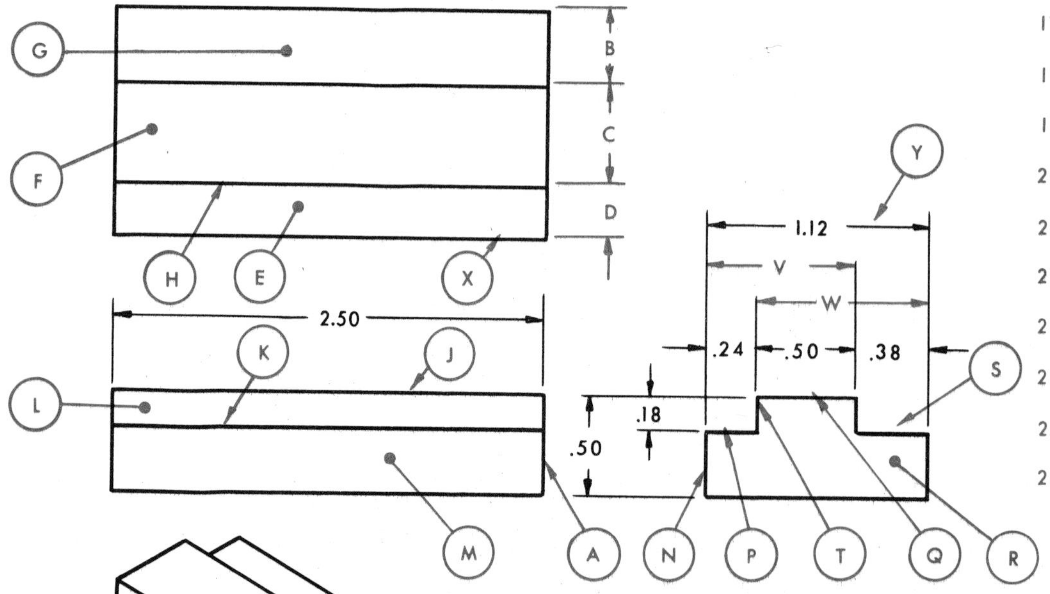

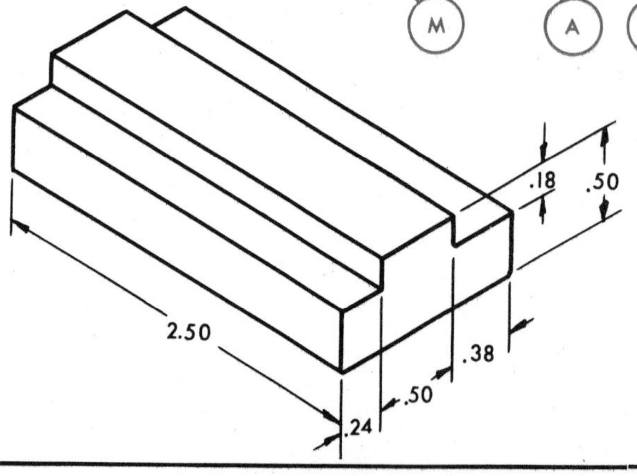

QUANTITY	2	
MATERIAL	M.S.	
SCALE	1/1	
DRAWN		DATE
COUNTER CLAMP BAR	A-3	

Drawing dimensions: 1.12, V, W, .24, .50, .38, .18, .50, 2.50, B, C, D

Isometric view dimensions: 2.50, .18, .50, .38, .50, .24

UNIT 3

HIDDEN LINES

In some objects, there are one or more hidden edges which cannot be seen from the outside of the piece. These hidden edges are called hidden lines. They are represented on a drawing by a series of small dashes.

The hidden lines should always start and end with a dash except when such a dash would form a continuation of a visible detail line. Dashes should always join at corners. Illustrations of these hidden line techniques are shown in figure 3-2.

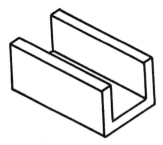

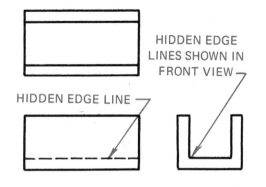

HIDDEN EDGE LINES SHOWN IN FRONT VIEW

HIDDEN EDGE LINE

Fig. 3-1   Hidden Lines.

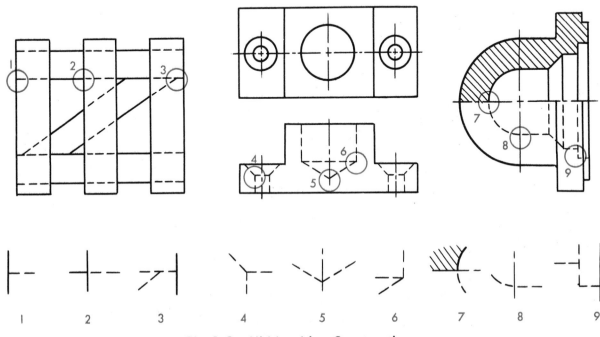

Fig. 3-2   Hidden Line Construction.

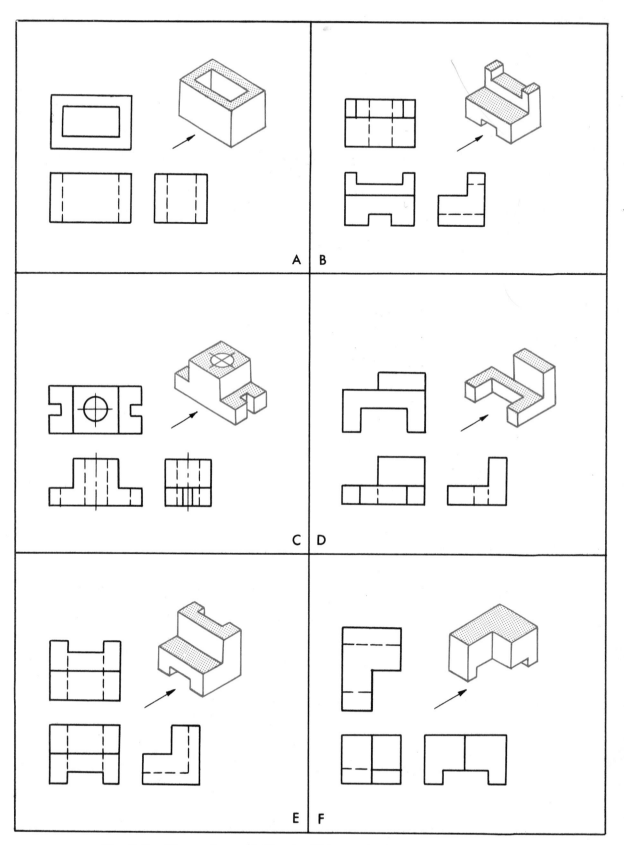

Fig. 3-3 Illustrations of Simple Objects Having Hidden Features.

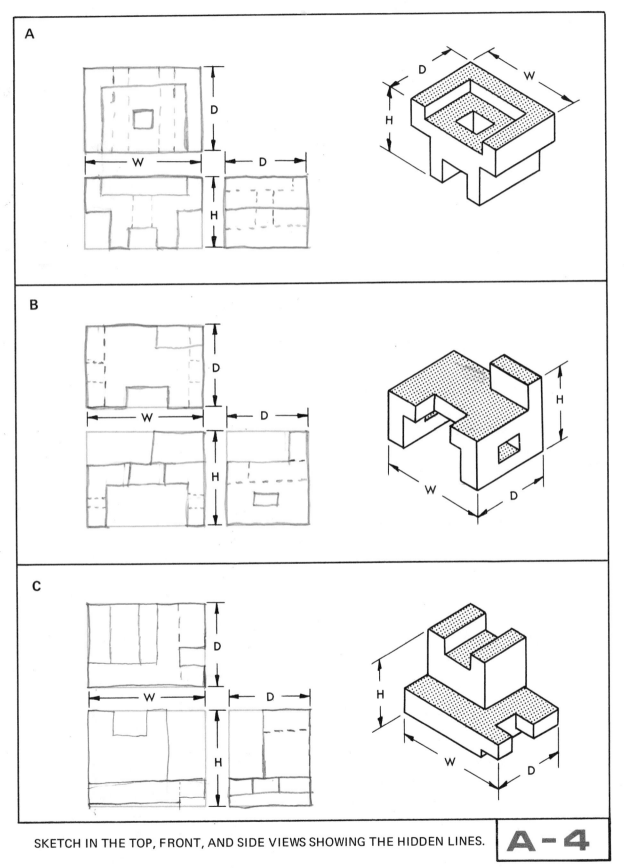

A

B

C

SKETCH IN THE TOP, FRONT, AND SIDE VIEWS SHOWING THE HIDDEN LINES.

A-4

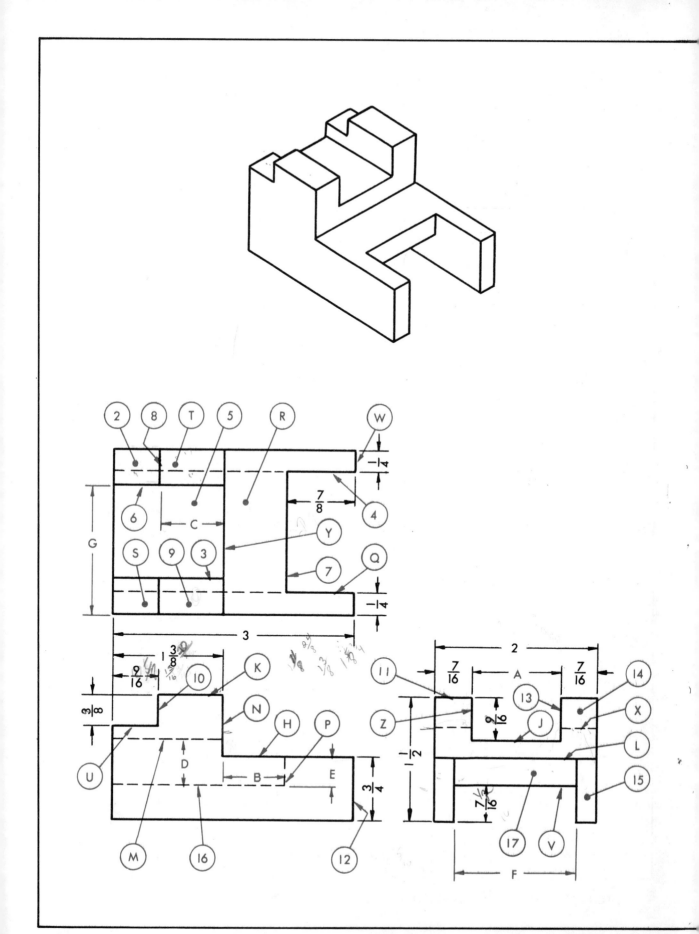

QUESTIONS

1. What is the name of the object?
2. What is the drawing number?
3. How many castings are required?
4. Of what material is the part made?
5. What is the overall length?
6. What is the overall height or thickness?
7. What is the overall width?
8. Calculate distances **A** through **G**.
9. What line in the top view represents surface (P) ?
10. What line in the side view represents surface (5) ?
11. What line in the side view represents surface (R) ?
12. What surfaces in the top view does line (K) of the front view represent?
13. What surface in the top view does line (M) in the front view represent?
14. What line in the side view represents the same surface represented by line (M) of the front view?
15. What kind, or type of line, is line (M) ?
16. What front view line does line (X) in the side view represent?
17. What surface in the side view does line (Y) in the top view represent?
18. What line in the front view does surface (15) in the side view represent?
19. What front view line represents surface (R) in the top view?
20. What surface in the side view represents line (N) of the front view?
21. What line in the side view represents surface (2) ?
22. What surface in the side view does line (P) represent?
23. What surface in the top view does line (11) represent?
24. What line in the side view does line (3) in the top view represent?
25. What line in the side view does line (16) in the front view represent?
26. What surface does (W) represent?
27. What type of dimensions are used on this drawing?

ANSWERS

1. Feed Hopper
2. A-5
3. 2
4. Cast Iron
5. 3
6. 1½
7. 2
8. A. 1⅛
   B. ¾
   C. 1³⁄₁₆
   D. ½
   E. ⁵⁄₁₆
   F. 1½
   G. 1⁹⁄₁₆
9. 7
10. J
11. H
12. 9+J
13. 5
14. J
15. Hidden
16. U
17. N
18. 12
19. H
20. 14
21. X
22. 17
23. 9
24. Z
25. V
26. 15
27. Fractional

QUANTITY	2 REQUIRED	
MATERIAL	C I cast iron	
SCALE	1/1	
DRAWN BY		DATE
FEED HOPPER		**A-5**

SLOPING SURFACES

If the surfaces of an object lie in either a horizontal or a vertical position, the surfaces appear in their true shapes in one of the three views, and these surfaces appear as a line in the other two views.

When a surface is sloped in only one direction, then that surface is not seen in its true shape in the top, front, or side views. It is, however, seen in two views as a distorted surface. On the third view it appears as a line.

The true length of surfaces A and B in figure 4-1 is seen in the front view only. In the top and side views, only the width of surfaces A and B appears in its true size. The length of these surfaces is foreshortened.

Where an inclined surface has important features that must be shown clearly and without distortion, an auxiliary or helper view must be used. These views will be discussed in detail later in the book.

MEASUREMENT OF ANGLES

Some objects do not have all of their straight lines drawn horizontally and vertically. The design of the part may require some lines to be drawn at an angle, either to each other, or to the horizontal or vertical.

The amount of this divergence, or obliqueness, of lines may be indicated by either an offset dimension or an angle dimension as shown in figure 4-2.

Angle dimensions are expressed in units of degree or units of degrees and minutes. On drawings the words *degree* and *minutes* are denoted by the symbols ° and '.

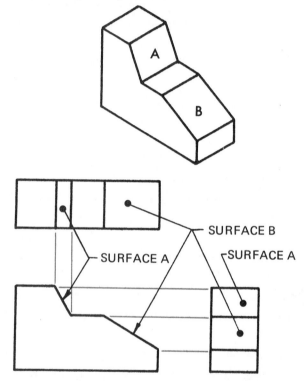

NOTE: THE TRUE SHAPE OF SURFACES A AND B DO NOT APPEAR ON THE TOP OR SIDE VIEWS.

Fig. 4-1   Sloping Surfaces.

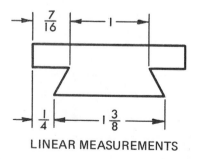

LINEAR MEASUREMENTS

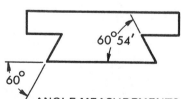

ANGLE MEASUREMENTS

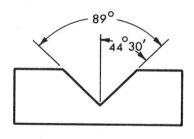

ANGLE MEASUREMENTS

Fig. 4-2   Dimensioning Angles.

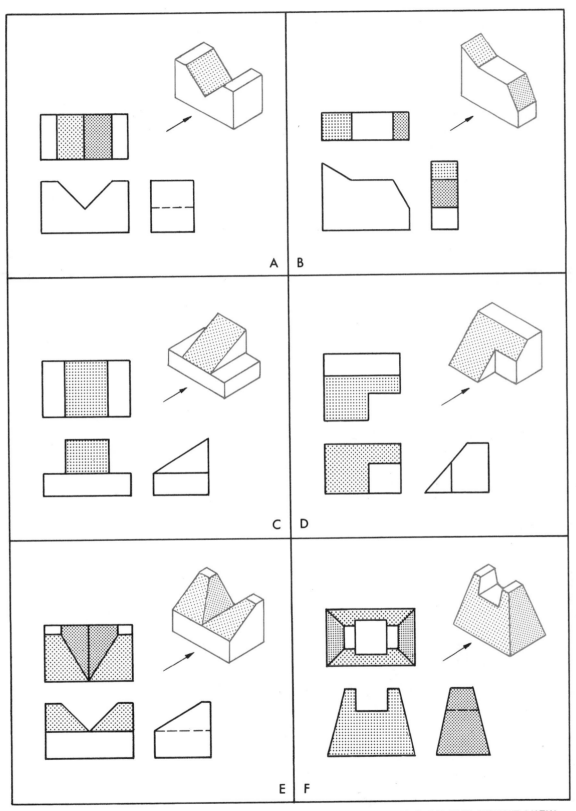

NOTE: ARROWS INDICATE DIRECTION OF SIGHT WHEN LOOKING AT THE FRONT VIEW.

Fig. 4-3   Illustrations of Simple Objects Having Sloped Surfaces.

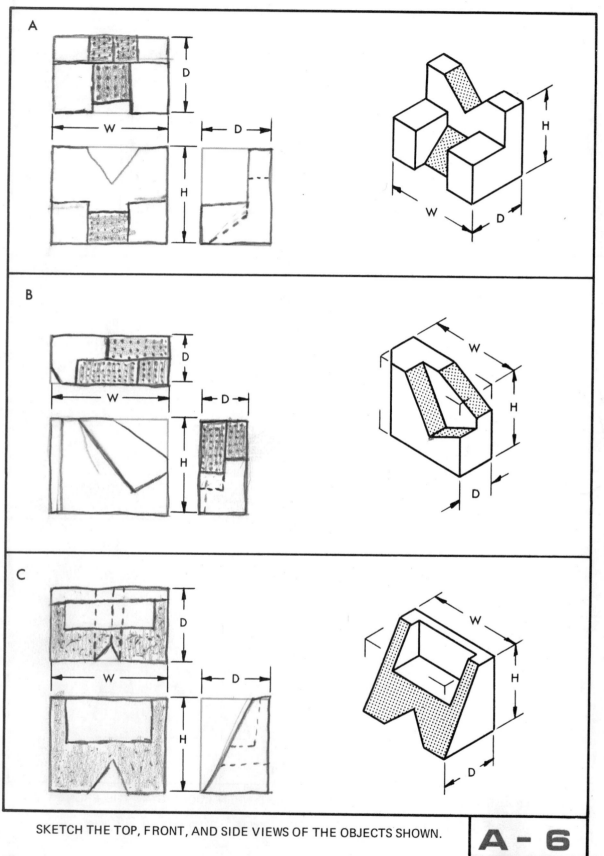

SKETCH THE TOP, FRONT, AND SIDE VIEWS OF THE OBJECTS SHOWN.

A-6

QUESTIONS

1. Calculate distances **A** to **G**.
2. At what angle is line ⑥ to the vertical?
3. At what angle is line ⑦ to the horizontal?
4. Locate surface ⑥ in the side view.
5. Locate surface ① in the side view.
6. Locate surface ⑥ in the top view.
7. What lines in the side view are represented by line ② in the front view?
8. Locate ⑱ in the top view.
9. Locate surface ⑨ in the side view.
10. Locate surface ⑫ in the front view.
11. Locate surface ③ in the top view.
12. What lines in the side view are represented by point ④ in the front view.
13. What line in the side view is line ⑯ in the top view?

14. Locate surface ⑩ in the side view.
15. Locate surface ⑩ in the front view.
16. Locate surface ⑫ in the side view.
17. Locate surface ⑮ in the side view.
18. What lines does point ④ represent in the top view?
19. Locate surface ⑥ in the top view.
20. Locate line ⑯ in the side view.
21. Locate line ㉕ in the top view.
22. What line in the front view is surface ⑮ in the top view?

ANSWERS

1 A	.18
B	.50
C	.70
D	1.50
E	.70
F	.86
G	.90
2	59° 20'
3	10°
4	19
5	26, 2
6	15
7	20+28
8	17
9	Not there
10	5
11	9
12	28 a 24
13	27
14	22
15	2
16	23
17	19
18	11 & 29
19	15
20	27
21	8
22	6

QUANTITY	4 REQD
MATERIAL	M.S.
SCALE	FULL SIZE
DRAWN	DATE

BASE PLATE **A-7**

19

UNIT 5

SCALE DRAWINGS

Many objects such as machine parts are too large to be drawn to full scale, so they must be drawn to a reduced scale. An example is a drawing of a lathe which probably would be drawn to 1/5 to 1/8 its actual size.

Frequently, objects such as small watch parts are drawn larger than their actual size so that their shape can be seen clearly. Such a drawing has been drawn to an enlarged scale. The gears in the wristwatch, for example, could be drawn to scale $10'' = 1''$.

The notation on a drawing *full size, half size,* or *one-fifth size* indicates the relationship between the size of the drawing and the actual size of the part. A half-size drawing is one-half as large as the part, but with full-scale dimensions.

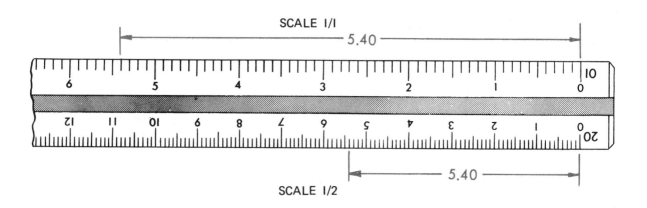

(A) DECIMALLY DIMENSIONED SCALE

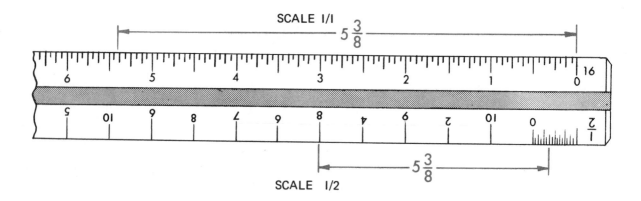

(B) FRACTIONALLY DIMENSIONED SCALE

Fig. 5-1 Engineering Drafting Scales.

DECIMALLY DIMENSIONED DRAWINGS		FRACTIONALLY DIMENSIONED DRAWINGS	
Proportion of Part to Drawing	Use Scale	Proportion of Part to Drawing	Use Scale
10/1	10	8/1	1 or 16
5/1	10	4/1	1 or 16
2/1	10	2/1	1 or 16
1/1	10	1/1	1 or 16
1/2	20	3/4	3/4
1/3	30	1/2	1/2
1/4	40	3/8	3/8
1/5	50	1/3	4″ = 1 ft.
1/6	60	1/4	1/4 or 3″ = 1 ft.
1/8	80	1/6	2″ = 1 ft.
1/10	10	1/8	1 1/2″ = 1 ft.
1/20	20	1/12	1″ = 1 ft.
etc.	etc.	1/16	3/4″ = 1 ft.
		1/24	1/2″ = 1 ft.
		1/32	3/8″ = 1 ft.
		1/48	1/4″ = 1 ft.
		1/64	3/16″ = 1 ft.
		1/96	1/8″ = 1 ft.
		1/128	3/32″ = 1 ft.

Fig. 5–2 Drawing Scales.

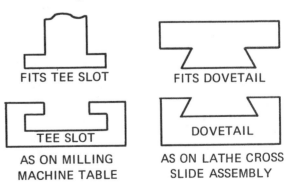

Fig. 5-3 Fillets and Rounds.

The scales used for decimally dimensioned and fractionally dimensioned drawings are shown in figure 5-2.

ROUNDS AND FILLETS

A round, or radius, is put on the outside of a piece to improve its appearance, and to avoid forming a sharp edge that might chip off under a sharp blow or cause interference. A fillet is additional metal allowed in the inner intersection of two surfaces to increase the strength of the object. A general note, such as *ROUNDS AND FILLETS .10 R* or *ROUNDS AND FILLETS .10 R UNLESS OTHERWISE SHOWN* is normally shown on the drawing instead of individual dimensions.

MACHINE SLOTS

Slots are used principally in machines to hold parts together. Two of the principal types are tee slots and dovetails which are shown in figure 5-4.

FITS TEE SLOT

FITS DOVETAIL

TEE SLOT

DOVETAIL

AS ON MILLING MACHINE TABLE

AS ON LATHE CROSS SLIDE ASSEMBLY

Fig. 5-4 Typical Machine Slots.

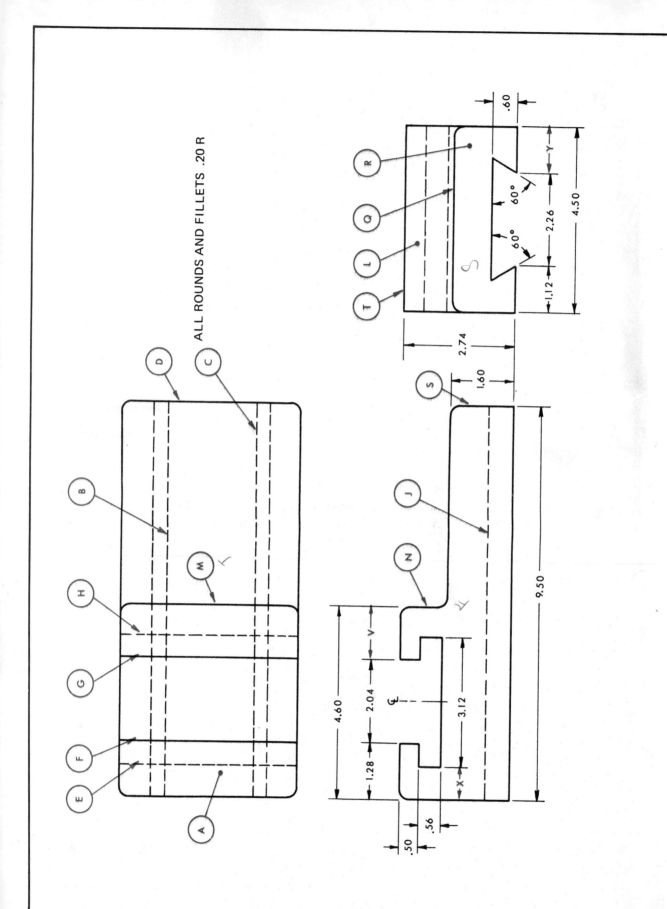

ALL ROUNDS AND FILLETS .20 R

QUESTIONS

1. In which view is the shape of the dovetail shown?

2. In which view is the shape of the T-slot shown?

3. In the top view how many rounds are shown?

4. In which view is a fillet shown?

5. What line in the top view represents surface (R) of the side view?

6. What line in the front view represents surface (R) ?

7. What line in the top view represents surface (L) of the side view?

8. What line in the front view represents surface (L) ?

9. What line in the side view represents surface (A) on the top view?

10. What dimension in the front view represents the width of surface (A) ?

11. What type of lines are (B) (J) (S) ?

12. How far apart are the two invisible edge lines of the side view?

13. What dimension indicates how far line (J) is from the base of the slide?

14. How wide is the opening in the dovetail?

15. What two lines in the top view indicate the opening of the dovetail?

16. At what angle to the horizontal is the dovetail cut?

17. In the side view, how far is the lower left edge of the dovetail from the left side of the piece?

18. What are the lengths of dimensions (Y) (V) (X) ?

19. To what depth into the piece is the dovetail cut?

20. How much material remains between the surface represented by line (Q) and the top of the dovetail after the cut has been taken?

21. What is the vertical distance from the surface represented by line (Q) to that represented by line (T) ?

22. What dimension represents the distance between lines (F) and (G) ?

23. What is the overall depth of the T-slot?

24. How thick is the metal at the sides of the 2.04 opening of the T-slot?

25. What is the width of the bottom of the T-slot?

26. What is the height of the opening of the bottom of the T-slot?

27. What is the horizontal distance from line (N) to line (S) ?

28. What type of dimensions are used on this drawing?

29. What is the unit of measurement of angles?

ANSWERS

1	RSide	16	30°
2	Front	17	1.12
3	6	18 Y	1.12
4	Front	V	1.28
5	D	X	.74
6	S	19	.60
7	M	20	1.00
8	N	21	1.14
9	T	22	2.04
10	1.28	23	1.06
11 B	Hidden	24	.50
J	Hidden	25	3.12
S	object	26	.56
12	.56	27	4.90
13	.60	28	Decimal
14	2.26	29	Degrees
15	B+C		

QUANTITY	2
MATERIAL	C I
SCALE	1/2
DRAWN	DATE

COMPOUND REST SLIDE

A-8

UNIT

6

CIRCULAR FEATURES

Typical parts with circular features are illustrated in figure 6-1. Note that the circular feature appears circular in one view only and that no line is used to indicate where a curved surface joins a flat surface. Hidden circles, like hidden flat surfaces, are represented on drawings by a hidden line.

Centerlines

A centerline is drawn as a thin broken line of long and short dashes, spaced alternately. They may be used to indicate center points, axes of cylindrical parts, and axes of symmetry. Solid centerlines are often used as a simplified drafting practice; however, the interrupted line is preferred. Centerlines should project for a short distance beyond the outline of the part or feature to which they refer. They may be extended for use as extension lines for dimensioning purposes, but in this case the extended portion is not broken.

In end views of circular features, the point of intersection of two centerlines is shown by two intersecting short dashes, except for very small circles where a solid unbroken line is recommended. These techniques are shown in figure 6-2.

Dimensioning of Cylindrical Features

Features shown as circles are normally dimensioned by one of the methods shown in

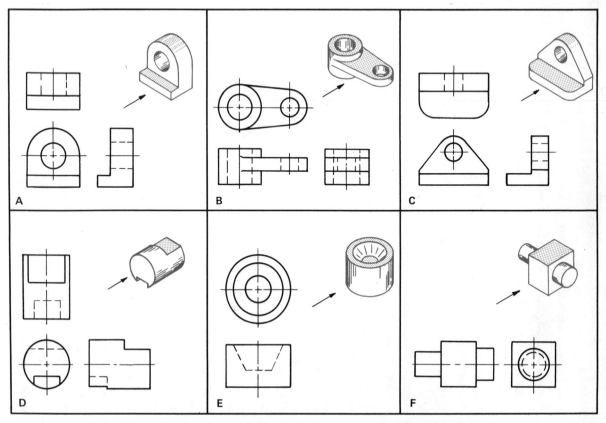

Fig. 6-1   Illustrations of Simple Objects Having Circular Features.

24

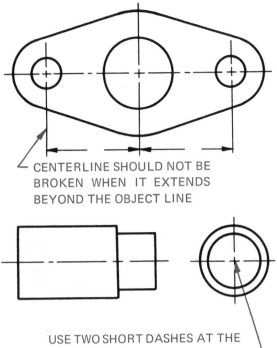

CENTERLINE SHOULD NOT BE
BROKEN WHEN IT EXTENDS
BEYOND THE OBJECT LINE

USE TWO SHORT DASHES AT THE
POINT OF INTERSECTION.

Fig. 6-2   Centerline Application.

figure 6-3. Where the diameters of a number of concentric cylinders are to be given, it may be more convenient to show them on the side view. The abbreviation *DIA* or the symbol $\emptyset$ is shown after the diametral dimension when an end view is not shown.

A circular arc is dimensioned by giving its radius. Approved methods for dimensioning arcs are shown in figure 6-4.

Dimensioning Cylindrical Holes

The preferred method of designating the size of small holes is to specify the diameter in inches with a leader, as shown in figure 6-5, or for larger holes, by means of one of the methods illustrated in figure 6-3. The abbreviation *DIA* or the symbol $\emptyset$ is given after the size of the hole when the leader is used. The note end of the leader terminates in a short horizontal bar. When more than one hole of a

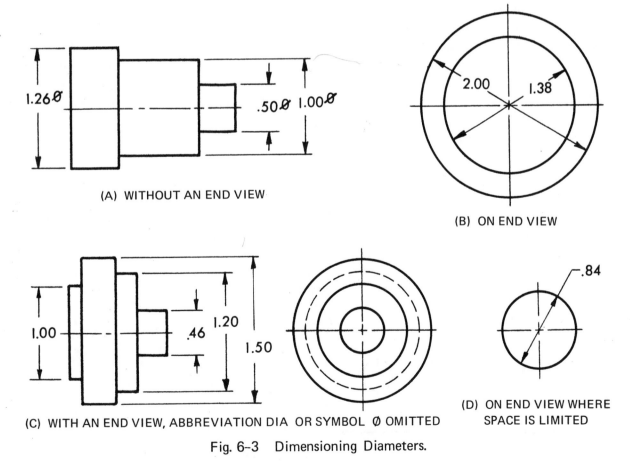

(A) WITHOUT AN END VIEW

(B) ON END VIEW

(C) WITH AN END VIEW, ABBREVIATION DIA OR SYMBOL $\emptyset$ OMITTED

(D) ON END VIEW WHERE SPACE IS LIMITED

Fig. 6-3   Dimensioning Diameters.

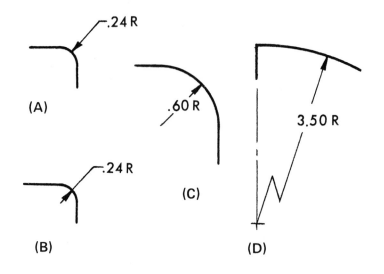

Fig. 6-4   Dimensioning Radii.

size is required, the number of holes is specified after the size. If a blind hole is required, the depth of the hole is included in the dimensioning note; otherwise, it is assumed that all holes shown are through holes. Terms such as drill, ream, bore, etc. should be avoided on drawings because the method of producing the hole should be left to the discretion of the man in the shop or the planning department.

DRILLING, REAMING, AND BORING

Drilling refers to the process of piercing a hole through a solid with a drill or the enlarging of a smaller hole. For some types of work holes must be drilled smooth, straight, and of an exact size. On others, the accuracy of location and size of hole is not so important.

When fairly accurate holes of uniform diameter are required, they are first drilled slightly undersize and then reamed. *Reaming* is the process of sizing a hole to a given diameter with a reamer in order to produce a hole which is round, smooth, and straight.

Boring is one of the more dependable methods of producing holes which are round and concentric. *Boring* refers to the enlarging of a hole by means of a boring tool. The application of boring differs from reaming in that the use of reamers is limited to the sizes of available reamers, while holes may be bored to any desired size.

The degree of accuracy to which a hole is to be machined is specified on the drawing and the method of producing the hole is left to the machinist.

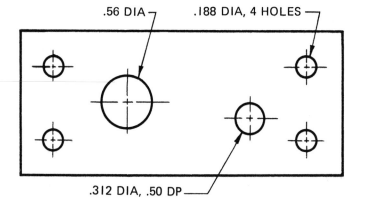

Fig. 6-5   Dimensioning Cylindrical Holes.

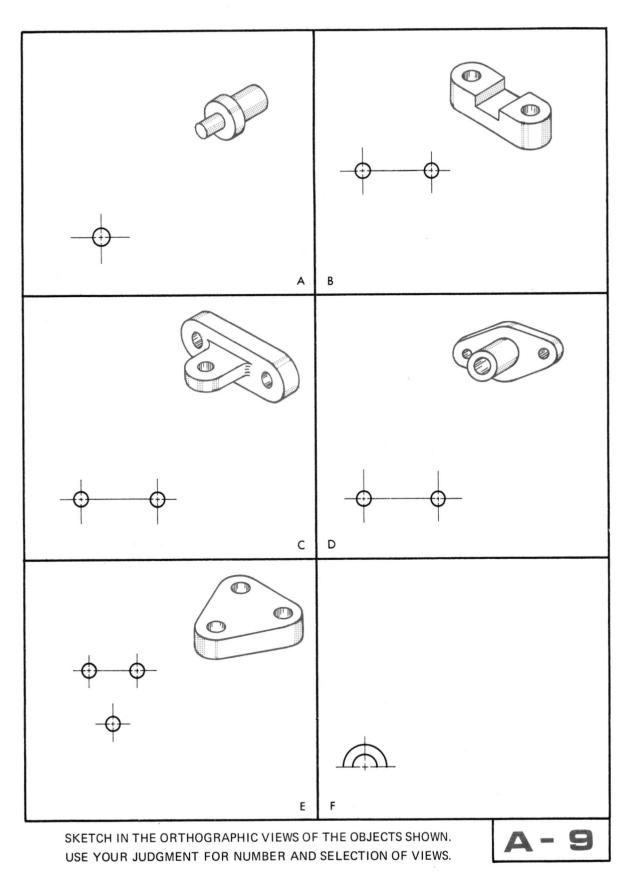

A   B

C   D

E   F

SKETCH IN THE ORTHOGRAPHIC VIEWS OF THE OBJECTS SHOWN.
USE YOUR JUDGMENT FOR NUMBER AND SELECTION OF VIEWS.

A- 9

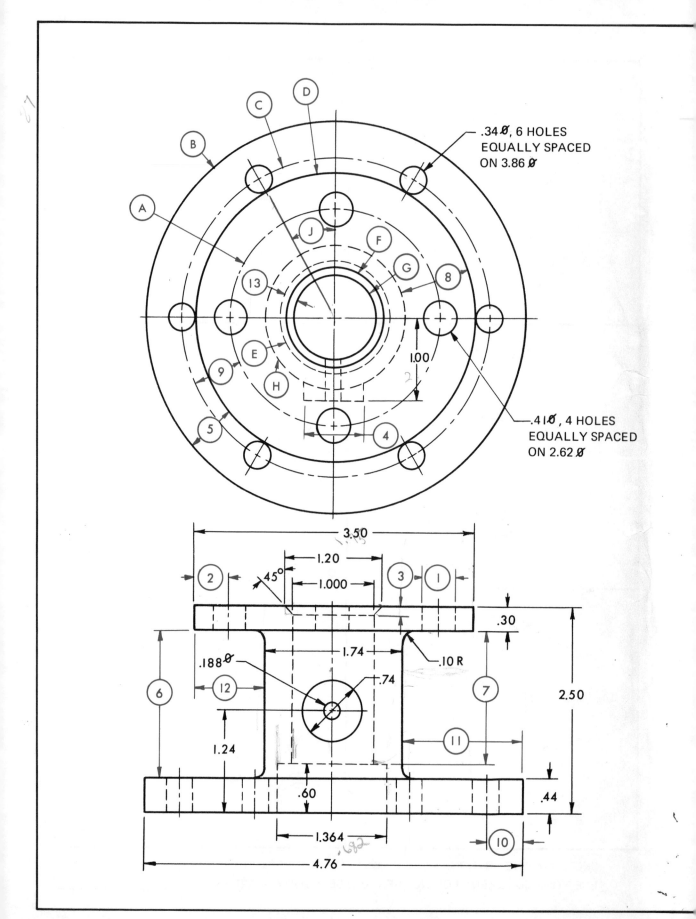

.34 Ø, 6 HOLES
EQUALLY SPACED
ON 3.86 Ø

.41Ø, 4 HOLES
EQUALLY SPACED
ON 2.62 Ø

1.00

3.50

1.20

1.000

45°

.30

1.74

.10 R

.188 Ø

.74

2.50

1.24

.60

.44

1.364

4.76

1. What are the diameters of circles Ⓐ to Ⓗ ?
2. How many holes are in the bottom flange?
3. How many holes are in the top flange?
4. What is the depth of the 1.00 dia. hole?
5. What is angle Ⓙ ?
6. What is the thickness of the largest flange?
7. What size bolts would be used for the top flange?
8. What size bolts would be used for the bottom flange?
9. Calculate distances ① to ⑬ .

ANSWERS

1 A 2.62
  B 4.76
  C 3.86
  D 3.50
  E 1.364
  F 1.20
  G 1.000
  H 1.74
2 6
3 4
4 1.90
5 30°
6 .44
7 3⁄8
8 3⁄16

9 ① .41
   ② .44
   ③ .10
   ④ .74
   ⑤ .63
   ⑥ 1.76
   ⑦ 1.60
   ⑧ .88
   ⑨ .62
   ⑩ .45
   ⑪ 1.51
   ⑫ .88
   ⑬ .82

QUANTITY	2	
MATERIAL	C I	
SCALE	1/1	
DRAWN		DATE
COUPLING		A-10

UNIT

7

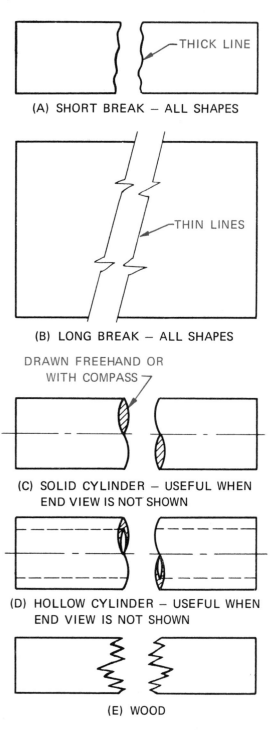

(A) SHORT BREAK — ALL SHAPES

THICK LINE

(B) LONG BREAK — ALL SHAPES

THIN LINES

DRAWN FREEHAND OR
WITH COMPASS

(C) SOLID CYLINDER — USEFUL WHEN
END VIEW IS NOT SHOWN

(D) HOLLOW CYLINDER — USEFUL WHEN
END VIEW IS NOT SHOWN

(E) WOOD

Fig. 7-1 Conventional Break Lines.

BREAK LINES

Break lines, as shown in figure 7-1, are used to shorten the view of long uniform sections or when only a partial view is required. Such lines are used on both detail and assembly drawings. The thin line with freehand zigzags is recommended for long breaks, the thick freehand line for short breaks, and the jagged line for wood parts. The special breaks shown for cylindrical and tubular parts are useful when an end view is not shown; otherwise, the thick break line is adequate.

NOT-TO-SCALE DIMENSIONS

When a dimension on a drawing is altered or when a part is broken so that its true length is not shown, it may be desirable to indicate that this dimension is not to scale. The letters *NTS* following the dimension, or a wavy line shown underneath the dimension as in figure 7-2, are two common methods used to indicate that the dimension is not drawn to scale.

MACHINING SYMBOLS [1]

In preparing working drawings of parts to be cast or forged, the draftsman must indicate the surfaces on the drawing which

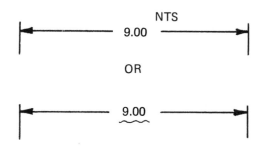

NTS
9.00

OR

9.00

Fig. 7-2 Indicating Dimensions That are Not to Scale.

require machining or finishing. This information is essential to the patternmaker who will provide extra metal on the casting to allow for the finishing process. Working surfaces such as bearings, pistons, and gears are typical of surfaces for which optimum performance may require control on the surface. Nonworking surfaces such as the walls of transmission cases, crankcases, or differential housings seldom require any surface control. In the mechanical field comparatively few surfaces require any control of smoothness or roughness beyond that afforded by the processes required to obtain the necessary dimension characteristics.

Smoothness and roughness are relative, that is, surfaces may be either smooth or rough for the purpose intended; what is smooth for one purpose may be rough for another purpose. Roughness or smoothness is a result of the processing method. The surface obtained from casting, forging or burnishing is the result of plastic deformation. If machined or ground, lapped or honed, the surface obtained is the result of tearing action of the cutting tools or abrasive grains.

Waviness

All surfaces, no matter how smooth they may appear, have a series of minute peaks and valleys which deviate in a more or less irregular fashion above and below a mean surface. This is known as waviness and results from such factors as machine or work deflection, vibration, chatter, heat treatment, and working strains.

Waviness height is rated in inches as the peak–to–valley distance. Waviness width is rated in inches as the spacing of successive wave peaks or successive wave valleys. When specified, values are the maximum permissible.

Roughness

Roughness superimposed on major peaks and valleys (waves) are irregularities of lesser magnitude. This roughness, which is caused mainly by the cutting edge of the tool and the tool feed, is given a *roughness height rating* which is the arithmetical average deviation from the roughness centerline expressed in microinches as measured normal to the centerline. *Roughness width* means the distance parallel to the nominal surface between successive peaks or ridges which constitute the predominant pattern of roughness. Roughness width is rated in inches.

Where it is necessary to indicate that a surface is to be machined without defining either the grade of roughness, or the process

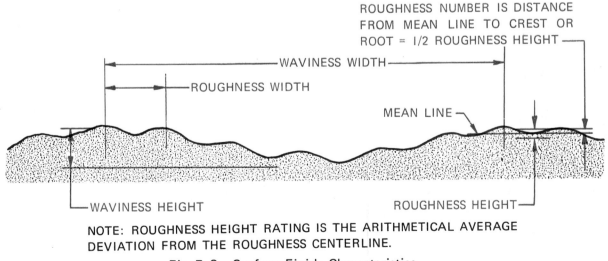

NOTE: ROUGHNESS HEIGHT RATING IS THE ARITHMETICAL AVERAGE DEVIATION FROM THE ROUGHNESS CENTERLINE.

Fig. 7-3 Surface Finish Characteristics.

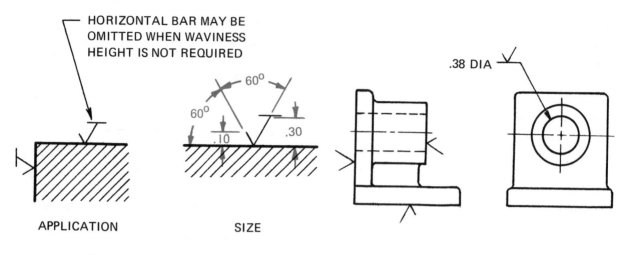

Fig. 7-4   Surface Symbol.

Fig. 7-5   Application of Surface Symbol.

to be used, the symbol shown in figure 7-4 should be applied as illustrated in figure 7-5. Where all the surfaces are to be machined, a general note such as *FINISH ALL OVER* or √ *ALL OVER* may be used and the symbols on the drawing may be omitted.

When it is necessary to indicate the grade of roughness, a roughness height rating is placed at the left of the long leg. The specification of only one rating indicates the maximum value and any lesser value is acceptable. The specification of two ratings indicates the minimum and maximum values, and anything lying within this range is acceptable. Maximum waviness height rating is placed above the horizontal extension. Any lesser rating is acceptable. Maximum waviness width rating is placed above the horizontal extension and to the right of the waviness height rating. Any lesser rating is acceptable. Machining symbols, like dimensions, are not normally duplicated. They should be placed on the same view as the dimensions that give the size or location of the surfaces concerned. Former surface symbols, as found on many drawings in use today are shown in figure 7-7. When called upon to make changes or revisions to a drawing already in existence, a draftsman must adhere to the drawing conventions shown on that drawing.

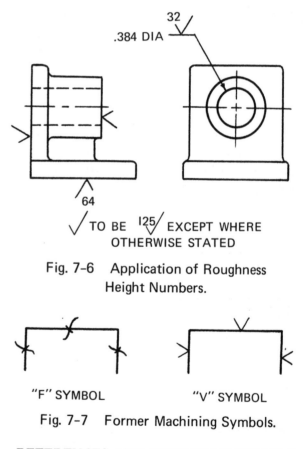

Fig. 7-6   Application of Roughness Height Numbers.

"F" SYMBOL   "V" SYMBOL

Fig. 7-7   Former Machining Symbols.

REFERENCES AND SOURCE MATERIALS

1. Extracted from *Surface Texture* (ASA B46.1 – 1962), with the permission of the publisher, The American Society of Mechanical Engineers, United Engineering Center, 345 East 47th Street, New York, N.Y. 10017.

Fig. 7-8 Surface Roughness Range for Common Production Methods.

PROCESS	ROUGHNESS HEIGHT (MICROINCHES)												
	2000	1000	500	250	125	63	32	16	8	4	2	1	0.5
FLAME–CUTTING													
SNAGGING													
SAWING													
PLANING, SHAPING													
DRILLING													
CHEMICAL MILLING													
ELEC. DISCHARGE MACH.													
MILLING													
BROACHING													
REAMING													
BORING, TURNING													
BARREL FINISHING													
ELECTROLYTIC GRINDING													
ROLLER BURNISHING													
GRINDING													
HONING													
POLISHING													
LAPPING													
SUPERFINISHING													
SAND CASTING													
HOT ROLLING													
FORGING													
PERM. MOLD CASTING													
INVESTMENT CASTING													
EXTRUDING													
COLD ROLLING, DRAWING													
DIE CASTING													

▬▬▬ AVERAGE APPLICATION

▭ LESS FREQUENT APPLICATION

THE RANGES SHOWN ABOVE ARE TYPICAL OF THE PROCESS LISTED: HIGHER OR LOWER VALUES MAY BE OBTAINED UNDER SPECIAL CONDITIONS.

Fig. 7-8 Surface Roughness Range for Common Production Methods. (Extracted from Surface Texture, ASA B46.1 – 1962, with the permission of the publisher, The American Society of Mechanical Engineers, United Engineering Center, 345 E. 47th St., New York, N.Y. 10017.)

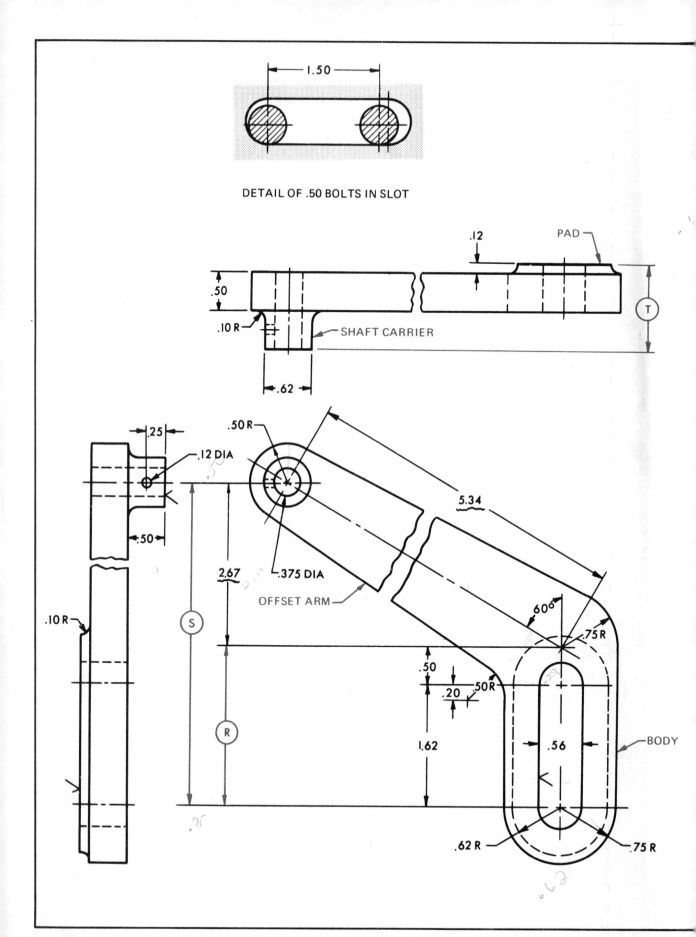

DETAIL OF .50 BOLTS IN SLOT

1.50

.12   PAD

.50

.10 R   SHAFT CARRIER

.62

T

.25

.12 DIA

.50 R

.375 DIA

OFFSET ARM

5.34

2.67

S

R

60°

.75 R

.50

.20

.50 R

1.62

.56

BODY

.10 R

.50

.62 R

.75 R

34

QUESTIONS

1. At what angle is the offset arm to the body of the piece?

2. What is the center-to-center measurement of the length of the offset arm?

3. What radius forms the upper end of the offset arm?

4. What radii form the lower end of the offset arm where it joins the body?

5. What is the width of the bolt slot in the body of the bracket?

6. What is the center-to-center length of this slot?

7. What is the end-to-end length of this slot?

8. What radius forms the ends of the pad?

9. What is the overall length of this pad?

10. What is the overall width of the pad?

11. What is the radius of the fillet between the pad and the edge of the piece?

12. What is the diameter of the shaft carrier body?

13. What diameter is the shaft carrier hole?

14. What is the distance from the face of the shaft carrier to the face of the pad?

15. What is the radius of the inside fillet between the arm and the body of the piece?

16. If .50″ bolts are used in holding the bracket to the machine base, what is the clearance on the sides of the slot?

17. If the center-to-center distance of two .50″ bolts which fit in the slot is 1.50″, how much play is there lengthwise in the slot?

18. What size oil hole is drilled in the shaft carrier?

19. How far is this hole drilled from the face of the shaft carrier?

20. How thick are the body and the pad together?

21. Calculate distances (R) (S) (T) .

22. What is the overall vertical height of the bracket?

23. How many dimensions are not to scale?

24. How many surfaces will be finished?

ANSWERS

1. _60°_
2. _5.34_
3. _.50 R_
4. _.75 R_
5. _.56″_
6. _1.62″_
7. _2.18″_
8. _.62 R_
9. _2.86″_
10. _1.24″_
11. _.10 R_
12. _.62″_
13. _.375″_
14. _1.12″_
15. _.50 R_
16. _.06″_
17. _.12″_
18. _.12 DIA_
19. _.25″_
20. _.62″_

21. (R) _2.12″_
    (S) _4.79″_
    (T) _1.12″_
22. _6.04″_
23. _2_
24. _3_

QUANTITY	75	
MATERIAL	C I	
SCALE	1/1	
DRAWN		DATE
OFFSET BRACKET		A-11

UNIT

8

SECTIONAL VIEWS [1]

Sectional views, commonly called sections, are used to show interior detail that is too complicated to be shown clearly by outside views and by the use of hidden lines. In assembly drawings, they also serve to indicate a difference in materials. A sectional view is obtained by supposing the nearest part of the object to be cut or broken away on an imaginary cutting plane. The exposed or cut surfaces are identified by section lining or crosshatching. Hidden lines and details behind the cutting plane line are usually omitted unless they

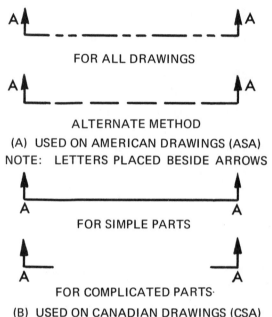

FOR ALL DRAWINGS

ALTERNATE METHOD

(A) USED ON AMERICAN DRAWINGS (ASA)
NOTE: LETTERS PLACED BESIDE ARROWS

FOR SIMPLE PARTS

FOR COMPLICATED PARTS·

(B) USED ON CANADIAN DRAWINGS (CSA)
NOTE — LETTERS PLACED BEHIND ARROWS

Fig. 8-2 Cutting Plane Lines.

are required for clarity or dimensioning. It should be understood that only in the sectional view is any part of the object shown as having been removed.

A sectional view frequently replaces one of the regular views. For example, a regular front view is replaced by a front view in section, as shown in figure 8-1.

The Cutting Plane Line

A cutting plane line is used to indicate where the imaginary cutting takes place. The position of the cutting plane is indicated, when necessary, on a view of the object or assembly by a cutting plane line as shown in figure 8-2. The ends of the cutting plane line are bent at 90 degrees and terminated by arrowheads to indicate the direction of sight for viewing the section. Cutting planes are not shown on sectional views. The cutting plane line may be

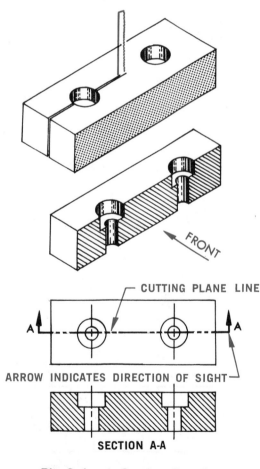

CUTTING PLANE LINE

ARROW INDICATES DIRECTION OF SIGHT

SECTION A-A

Fig. 8-1 A Section Drawing.

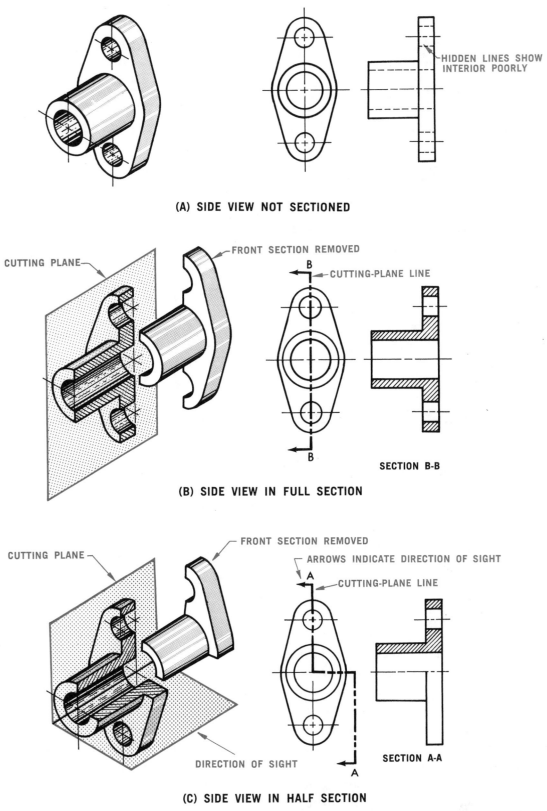

(A) SIDE VIEW NOT SECTIONED

HIDDEN LINES SHOW INTERIOR POORLY

CUTTING PLANE

FRONT SECTION REMOVED

B — CUTTING-PLANE LINE

SECTION B-B

(B) SIDE VIEW IN FULL SECTION

CUTTING PLANE

FRONT SECTION REMOVED

ARROWS INDICATE DIRECTION OF SIGHT

A — CUTTING-PLANE LINE

DIRECTION OF SIGHT

SECTION A-A

(C) SIDE VIEW IN HALF SECTION

Fig. 8-3   Full and Half Sections.

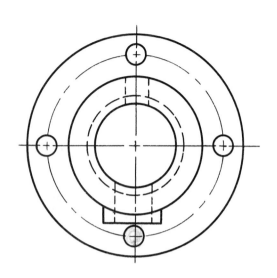

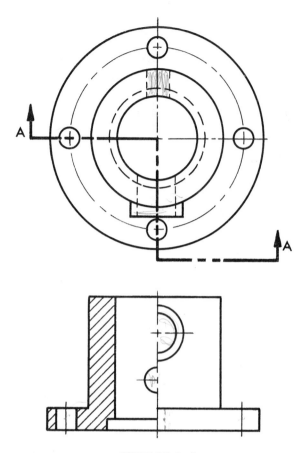

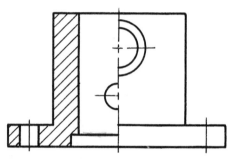

SECTION A–A

Letters, subtitle, and cutting plane line
used when more than one section view
appears on a drawing or when it makes
the drawing clearer

Letters, subtitle and cutting plane line
may be omitted when it corresponds with
the centerline of the part and when there
is only one section view on the drawing

Fig. 8-4   Identification of Cutting Plane and Section View.

omitted when it corresponds to the centerline
of the part or when only one sectional view
appears on a drawing.

If two or more sections appear on the
same drawing, the cutting plane lines are identified by two identical large, single-stroke,
Gothic letters. One letter is placed at each end
of the line, (a) adjacent to the bends at the end
of the cutting plane line for ASA drawings or
(b) behind the arrowhead so that the arrow
points away from the letter for CSA drawings.
Sectional view subtitles are given when identification letters are used and appear directly
below the view, incorporating the letters at

each end of the cutting plane line thus: *SEC-
TION A–A,* or abbreviated, *SECT. A–A.*

Section Lining

Section lining indicates the surface that
has been cut and makes it stand out clearly.
Section lines usually consist of thin parallel
lines, as shown in figure 8-4, drawn at an angle
of approximately 45 degrees to the principal
edges or axis of the part.

When it is desirable to indicate differ-
ences in materials, other symbolic section
lines are used, such as those shown in figure

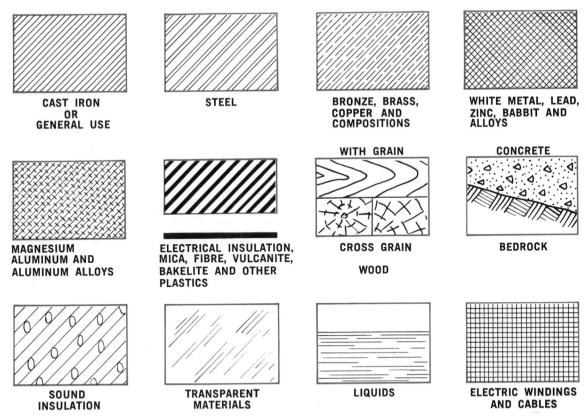

CAST IRON OR GENERAL USE

STEEL

BRONZE, BRASS, COPPER AND COMPOSITIONS

WHITE METAL, LEAD, ZINC, BABBIT AND ALLOYS

MAGNESIUM ALUMINUM AND ALUMINUM ALLOYS

ELECTRICAL INSULATION, MICA, FIBRE, VULCANITE, BAKELITE AND OTHER PLASTICS

WITH GRAIN

CROSS GRAIN

WOOD

CONCRETE

BEDROCK

SOUND INSULATION

TRANSPARENT MATERIALS

LIQUIDS

ELECTRIC WINDINGS AND CABLES

Fig. 8-5 Symbolic Section Lining.

8-5. If the part shape would cause section lines to be parallel, or nearly so, to one of the sides or features of the part, some angle other than 45 degrees is chosen.

The spacing of the hatching lines is reasonably uniform to give a good appearance to the drawing. The pitch, or distance, between lines varies between .05 inch to .10 inch, depending on the size of the area to be sectioned. In all sections of a single component, section lining is similar in direction and spacing.

Wood and concrete are the only two materials generally shown symbolically. When wood symbols are used, the direction of the grain is shown.

TYPES OF SECTIONS

Full Sections

When the cutting plane extends entirely through the object in a straight line and the front half of the object is theoretically removed, a *full section* is obtained. This type of section is used for both detail and assembly drawings. When the section is on an axis of symmetry, it is not necessary to indicate its location. However, it may be identified and indicated in the normal manner to increase clarity, if so desired.

Half Sections

A symmetrical object or assembly may be drawn as a *half section*, showing one half up to the centerline in section and the other half in full view. A normal centerline is used on the section view.

The half section drawing is not desirable where the dimensioning of internal diameters is required to avoid the adding of hidden lines to the portion showing the external features. This type of section is used mostly for assembly drawings where internal and external

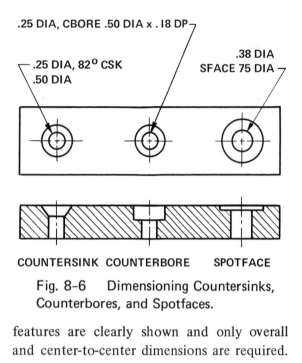

.25 DIA, CBORE .50 DIA x .18 DP

.25 DIA, 82° CSK
.50 DIA

.38 DIA
SFACE 75 DIA

COUNTERSINK   COUNTERBORE   SPOTFACE

Fig. 8-6   Dimensioning Countersinks,
Counterbores, and Spotfaces.

features are clearly shown and only overall
and center-to-center dimensions are required.

COUNTERSINKS, COUNTERBORES, AND SPOTFACES

A countersunk hole is a conical depression cut in a piece to receive a countersunk
type of flathead screw or rivet as illustrated in
figure 8-6. The size is usually shown in the
form of a note listing the diameter of the hole
first, followed by the angle, the abbreviation
CSK and lastly the diameter of the countersink. A counterbored hole is one which has
been machined larger to a given depth to receive a fillister, hex-head, or similar type of
bolt head. Counterbores are specified by a
note giving the diameter of the hole first, followed by the abbreviation *CBORE* and the
dimensions for the diameter and depth of the
counterbore. The counterbore and depth may
also be indicated by direct dimensioning. A
spotface is an area where the surface is machined just enough to provide a level seating
surface for a bolt head, nut or washer. A spotface is specified by a note listing the diameter
of the hole first, followed by the abbreviation *SFACE* and the dimension for the diameter. The depth of the spotface is not normally given.

REFERENCES AND SOURCE MATERIALS

1. Canadian Standards Association, *Bulletin* 78.1 (1967).

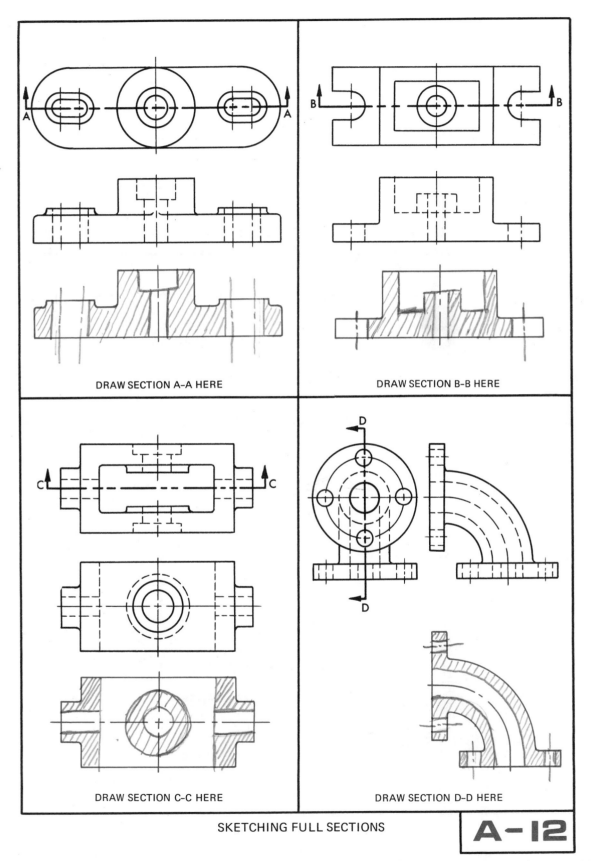

DRAW SECTION A-A HERE

DRAW SECTION B-B HERE

DRAW SECTION C-C HERE

DRAW SECTION D-D HERE

SKETCHING FULL SECTIONS

A-12

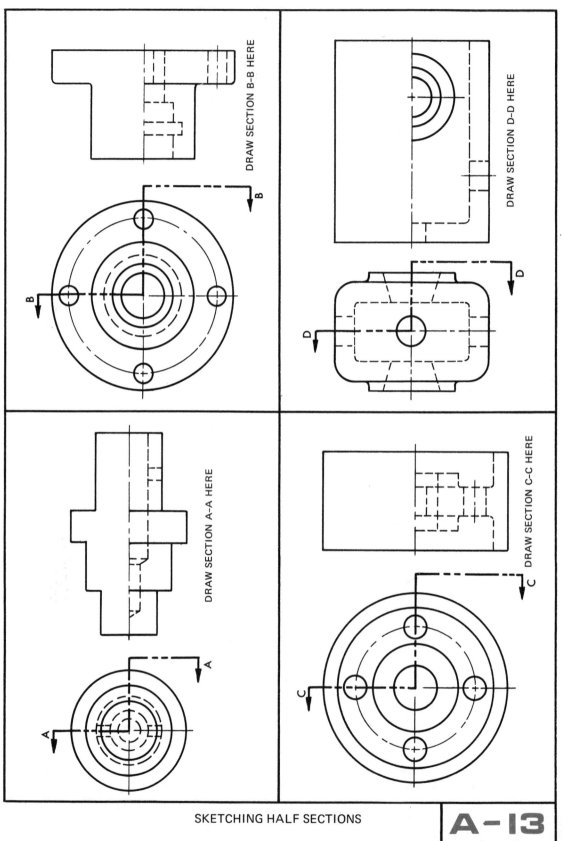

DRAW SECTION B-B HERE

DRAW SECTION D-D HERE

DRAW SECTION A-A HERE

DRAW SECTION C-C HERE

SKETCHING HALF SECTIONS

A-13

QUESTIONS

1. What is the overall length?
2. What is the overall height?
3. What is the center-to-center distance of the .19 dia. holes?
4. How many holes are in the bracket?
5. At what angle to the vertical is the dovetail slot?
6. How many different surfaces require finishing?
7. What type of lines in the top view represent the dovetail?
8. What type of lines in the front view represent the dovetail?
9. How wide is the opening in the dovetail?
10. How deep is the dovetail?
11. How deep are the .19 dia. holes?
12. Calculate distances (A) to (S).

ANSWERS

1 ——————— 11 ——————— (K) ———————
2 ——————— 12 (A) ——————— (L) ———————
3 ——————— (B) ——————— (M) ———————
4 ——————— (C) ——————— (N) ———————
5 ——————— (D) ——————— (P) ———————
6 ——————— (E) ——————— (Q) ———————
7 ——————— (F) ——————— (R) ———————
8 ——————— (G) ——————— (S) ———————
9 ——————— (H) ———————
10 ——————— (J) ———————

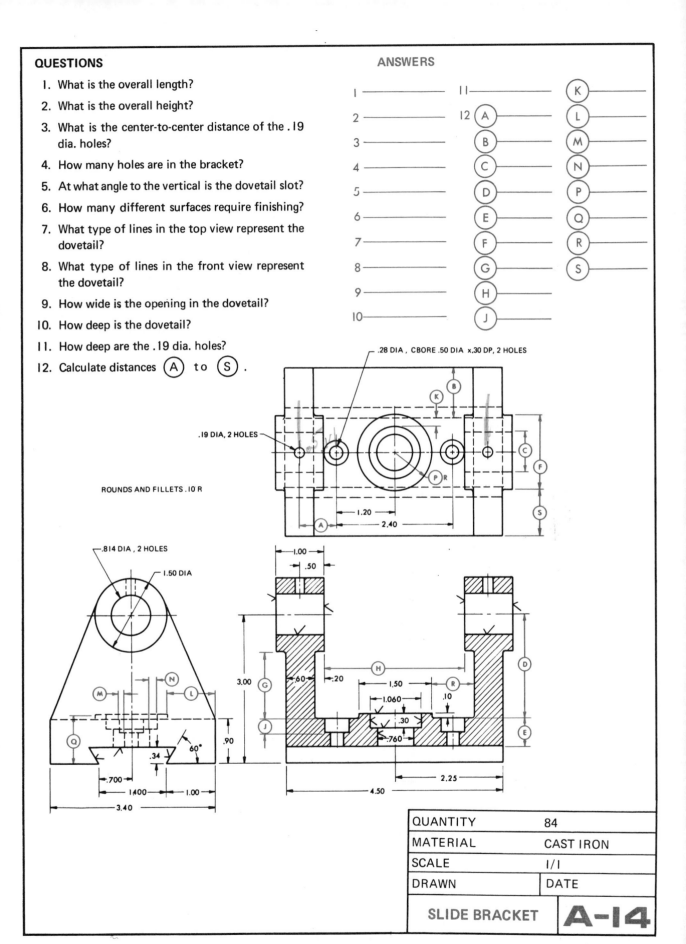

.28 DIA , CBORE .50 DIA x.30 DP, 2 HOLES

.19 DIA, 2 HOLES

ROUNDS AND FILLETS .10 R

1.20
2.40

.814 DIA , 2 HOLES
1.50 DIA

3.00
.90
60°
.34
.700
1.400
1.00
3.40

1.00
.50
.60
.20
1.50
1.060
.30
.760
.10
3.00
2.25
4.50

QUANTITY	84
MATERIAL	CAST IRON
SCALE	1/1
DRAWN	DATE

SLIDE BRACKET **A-14**

43

UNIT 9

ONE-AND TWO-VIEW DRAWINGS [1]

Except for complex objects of irregular shapes, it is seldom necessary to draw more than three views, and for simple parts one- or two-view drawings will often suffice.

In one-view drawings the third dimension may be expressed by a note or by descriptive words or abbreviations, such as *diameter,* or *hexagon across flats.* Square sections may be indicated by light crossed diagonal lines. The centerline through the piece indicates that it is symmetrical. Frequently, the draftsman will decide that only two views are necessary to explain the shape of an object fully. One or two views are usually sufficient to show the shape of cylindrical objects adequately.

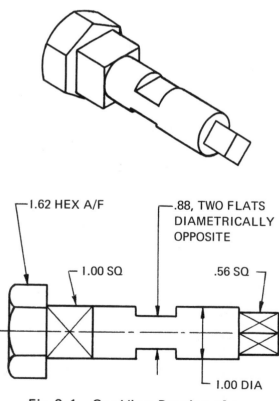

Fig. 9-1   One-View Drawing of a Turned Part.

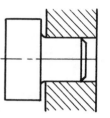

Part cannot fit flush in hole because of shoulder

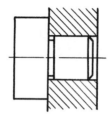

Same part with neck added permits part to fit flush

(A) CHAMFER AND NECK APPLICATION

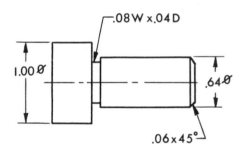

NOTE: THIS METHOD OF DIMENSIONING FOR 45° CHAMFER ONLY

(B) DIMENSIONING CHAMFER AND NECK

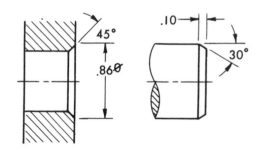

(C) DIMENSIONING CHAMFERS

Fig. 9-2   Chamfers and Necks.

CHAMFERING

The process of chamfering, that is, cutting away the inside or outside piece, is done to facilitate assembly. The recommended method of dimensioning a chamfer is to give an angle and a length, or an angle and a diameter. For angles of 45 degrees only, a note form may be used. This method is permissible only with 45-degree angles because the size may apply to either the longitudinal or radial dimension. Chamfers are never measured along the angular surface.

NECKING

The operation of necking or grooving, that is, cutting a recess in a diameter, is done to permit two parts to come together as illustrated in figure 9-2. It is indicated on a drawing by a note listing the width first and then the depth. The word *neck* need not be shown.

Where the size of the neck is unimportant, the dimension may be left off the drawing.

DETAIL DRAWING

Details of parts may be shown on separate sheets, or they may be grouped together on one or more large sheets. Often the details of parts are grouped according to the department in which they are made. Metal parts to be fabricated in the machine shop may appear on one detail sheet while parts to be made in the wood shop may be grouped on another.

REFERENCES AND SOURCE MATERIALS

1. Canadian Standards Association, *Bulletin* 78.1 (1967).

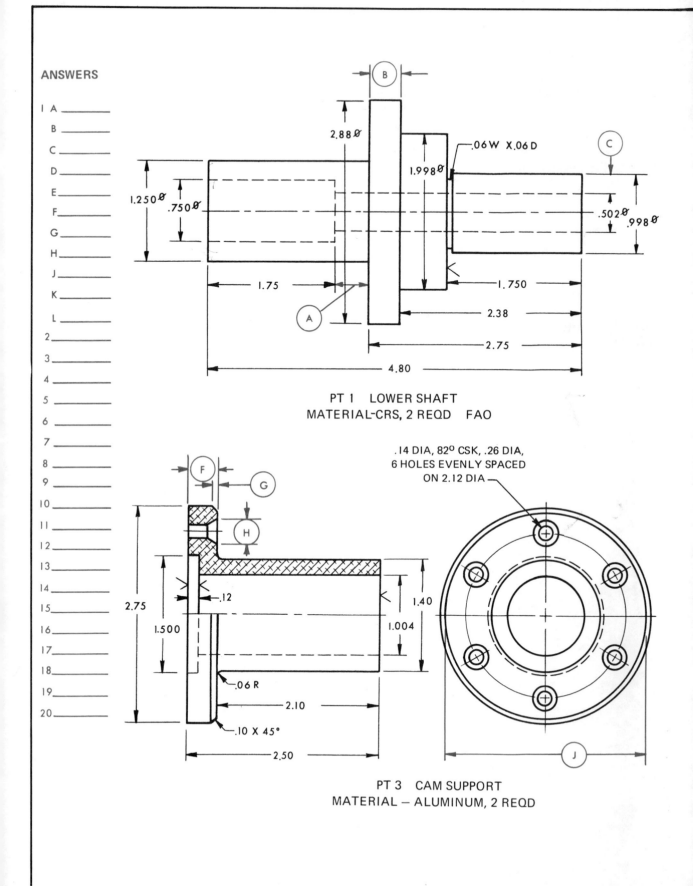

PT 1   LOWER SHAFT
MATERIAL-CRS, 2 REQD   FAO

PT 3   CAM SUPPORT
MATERIAL — ALUMINUM, 2 REQD

46

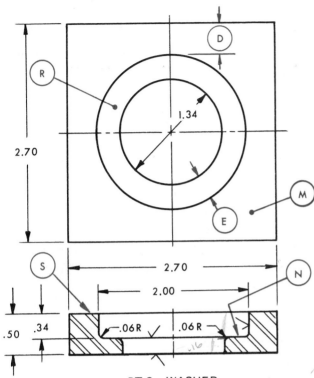

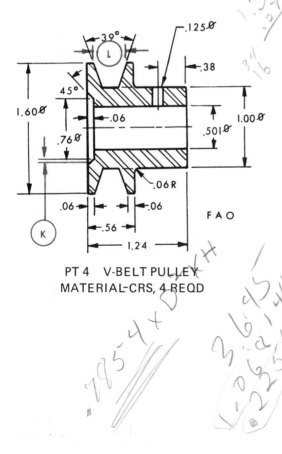

PT 2   WASHER
MATERIAL – MACH ST, 4 REQD

PT 4   V-BELT PULLEY
MATERIAL-CRS, 4 REQD

FAO

QUESTIONS

1. Calculate distances (A) to (L).

2. How many through holes are there on the four different parts?

3. What is the total number of parts required?

4. How many chamfers are shown on the drawing?

Refer to Part 1

5. What is the length of the portion that is .998" in diameter?

6. How many visible circles would be seen if the left end view were drawn?

7. How many hidden circles would be seen if the right end view were drawn?

Refer to Part 2

8. What surface does (R) represent in the front view?

9. What surface does (S) represent in the top view?

10. If the part that passes through the washer is 1.30" in diameter, what is the clearance per side between the two parts?

11. How many sharp edges are broken?

Refer to Part 3

12. How many degrees apart are the .14 dia. holes?

13. Is the centerline of the countersunk holes in the center of the flange?

14. What operation is performed to allow the head of the mounting screws to rest flush with the flange?

15. What type of section view is shown?

16. What is the amount and degree of chamfer?

Refer to Part 4

17. How deep is the .125 dia. hole?

18. How deep is the belt groove?

19. What does FAO mean?

20. What type of section view is shown?

SCALE	1/1
DRAWN	
	DATE
HANGER DETAILS	A-15

UNIT
10

TOLERANCES AND ALLOWANCES [1]

In the 6,000 years of the history of engineering drawing as a means for the communication of engineering information, it seems inconceivable that such an elementary practice as the tolerancing of dimensions, which we take so much for granted today, was introduced for the first time only about 60 years ago.

Apparently, engineers and workmen came to the realization only in a very gradual manner that *exact* dimensions and shapes could not be attained in the shaping of physical objects, such as the manufacture of materials and products. The skilled handicraftsman of old prided himself on being able to work to

exact dimensions. What he really meant was that he dimensioned objects with a degree of accuracy greater than that with which he could measure them. The use of modern measuring instruments would have shown the deviations from the sizes which he called exact.

As soon as it was realized that variations in the sizes of parts had always been present, that such variations could be restricted but not avoided, and also that slight variation in the size which a part was originally intended to have could be tolerated without its correct functioning being impaired, it was evident that interchangeable parts need not be identical parts, but that it would be sufficient if the significant sizes which controlled their fits lay

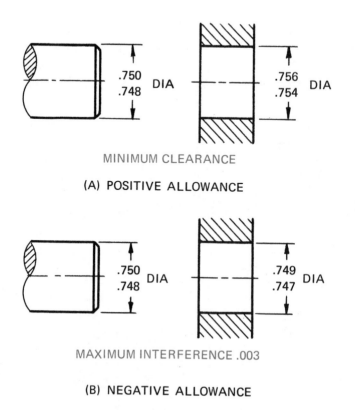

MINIMUM CLEARANCE

(A) POSITIVE ALLOWANCE

MAXIMUM INTERFERENCE .003

(B) NEGATIVE ALLOWANCE

Fig. 10-1   Allowances.

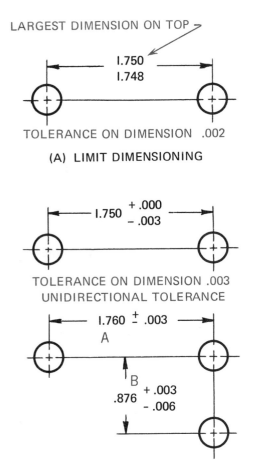

LARGEST DIMENSION ON TOP

1.750
1.748

TOLERANCE ON DIMENSION .002

(A) LIMIT DIMENSIONING

1.750 +.000
       −.003

TOLERANCE ON DIMENSION .003
UNIDIRECTIONAL TOLERANCE

1.760 ± .003
A

B
.876 +.003
      −.006

TOLERANCE ON DIMENSION (A) .006
TOLERANCE ON DIMENSION (B) .009

BILATERAL TOLERANCES

(B) PLUS AND MINUS TOLERANCES

Fig. 10-2 Tolerancing Methods.

between definite limits. Accordingly, the problem of interchangeable manufacture developed from the making of parts to a supposedly exact size to the holding of parts between two limiting sizes, lying so closely together that any intermediate size would be acceptable.

The development of the conception of limits meant essentially that an exactly defined *basic* condition, expressed by one numerical value or specification, was replaced by two limiting conditions, and any result lying between these two limits was acceptable. One standard level was replaced by two limiting levels enclosing a zone of acceptability, or tolerance, and a workable scheme of interchangeable manufacture, indispensable to mass production methods, thus became established.

Definitions [2]

Tolerances. The tolerance on a dimension is the total permissible variation in its size. The tolerance is the difference between the limits of size.

Limits of Size. The limits are the maximum and minimum sizes permitted for a specific dimension.

Allowance. An allowance is the intentional difference in correlated dimensions of mating parts. It is the minimum clearance (positive allowance) or maximum interference (negative allowance) between such parts.

All dimensions on a drawing have tolerances. Some dimensions must be more exact than other dimensions and as a result have smaller tolerances. General practice is to allow a deviation in size of ± .01 for all dimensions shown to two decimal places, ± .001 for all dimensions shown to three decimal places and ± 1/64 for all fractional dimensions. This tolerance is usually shown in the form of a general note on the drawing. Example: *"Except where stated otherwise, tolerances on dimensions ± .01."*

Where dimensions require a greater accuracy than that provided by the general note, then individual tolerances or limits will be shown for that dimension. For example: .346 ± .001 or .345/.347.

Tolerancing Methods

Dimensional tolerances are expressed in either of two ways: limit dimensioning or plus and minus tolerancing.

Limit Dimensioning. In the limiting dimension method, only the maximum and minimum dimensions are specified.

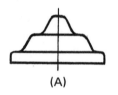

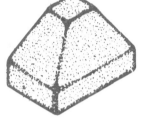

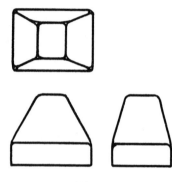

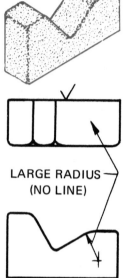

(A)

(B)

LARGE RADIUS
(NO LINE)

(C)

ROUNDED AND FILLETED INTERSECTIONS

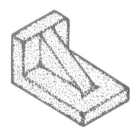

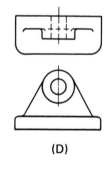

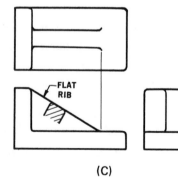

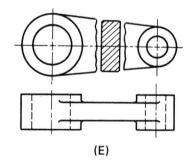

FLAT
RIB

(C)

(D)

(E)

RUNOUTS

Fig. 10-3 Rounded and Filleted Intersections.

Plus and Minus Tolerancing. For this method, the dimension of the specified size is given first, and is followed by a plus and minus expression of tolerance.

INTERSECTION OF UNFINISHED SURFACE

The intersection of unfinished surfaces that are rounded or filleted at the point of theoretical intersection may be indicated conventionally by a line coinciding with the theoretical point of intersection. The need for this convention is shown by the examples shown in figure 10-3. For a large radius such as shown in figure 10-3C, no line is drawn.

Members such as ribs and arms that blend into other features end in curves called runouts.

REFERENCES AND SOURCE MATERIALS

1. F.L. Spalding, "The Development of Standards for Dimensioning and Tolerancing," *Graphic Science,* 8 No. 2, (1966).

2. Extracted from *USA Standard Drafting Practices: Dimensioning and Tolerancing for Engineering Drawings* (USASI Y14.5 — (1966), with the permission of the publisher, The American Society of Mechanical Engineering Center, 345 East 47th Street, New York, N.Y. 10017.

NOTE: EXCEPT WHERE NOTED —
TOLERANCES ON TWO DECIMAL DIMENSIONS ± .02
TOLERANCES ON THREE DECIMAL DIMENSIONS ± .005
ROUNDS AND FILLETS .10 R

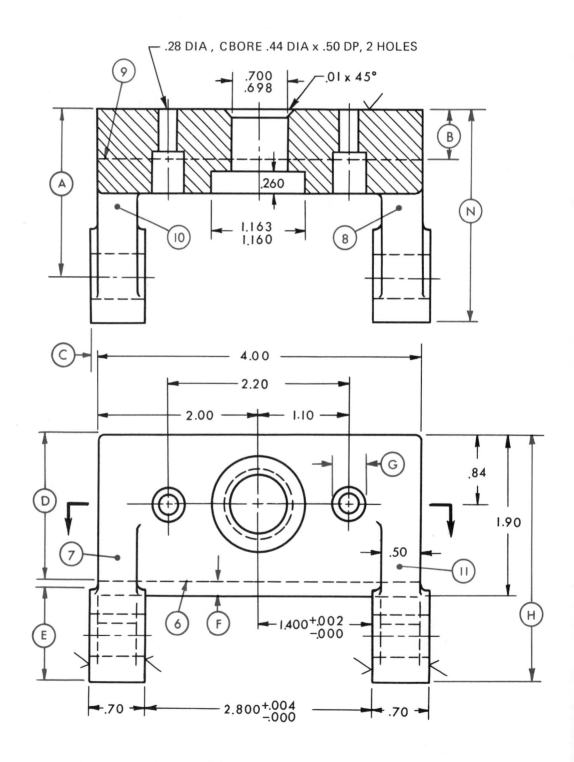

QUESTIONS

1. How many surfaces are to be finished?

2. Except where noted otherwise, what is the tolerance on (A) two-decimal dimensions, (B) three-decimal dimensions?

3. What is the tolerance on the .500 dia. holes?

4. What are the limit dimensions for the 1.600 dimension shown on the side view?

5. What are the limit dimensions for the 1.00 dimension shown on the side view?

6. What is the maximum permissible distance between the centers of the .28 dia. holes?

7. Express the .500 dia. hole as a plus-and-minus tolerance dimension.

8. What is the maximum tolerance placed on dimension (H) ?

9. What are the limit dimensions for the .260 dimension shown on the top view?

10. Locate surfaces (4) on the top view.

11. How many surface (5) s are there?

12. Locate line (3) in the top view.

13. Locate line (6) in the side view.

14. What surfaces in the front view indicate line (4) in the side view?

15. Calculate distances (A) to (N).

ANSWERS

1. _7_

2A. _± .02_

B. _± .005_

3. _.495 - .505_

4. _____

5. _____

6. _____

7. _____

8. _____

9. _____

10. _____

11. _____

12. _____

13. _____

14. _____

15 (A) _____

(B) _____

(C) _____

(D) _____

(E) _____

(F) _____

(G) _____

(H) _____

(J) _____

(K) _____

(L) _____

(M) _____

(N) _____

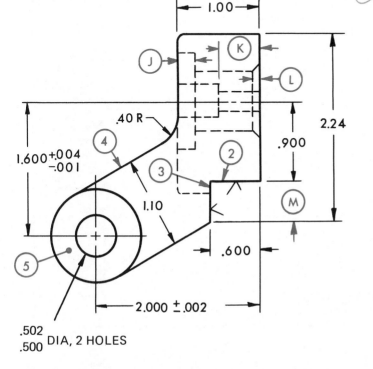

.502
.500 DIA, 2 HOLES

QUANTITY	2	
MATERIAL	C1	
SCALE	1/1	
DRAWN		DATE
BRACKET		**A-16**

THREAD REPRESENTATION [1]

True representation of a screw thread is seldom provided on working drawings because of the cost of drawing it. Three types of conventions are in general use for screw thread representation: pictorial, schematic, and simplified presentation. Simplified representation is used whenever it will clearly portray the requirements. Schematic and pictorial representations require more drafting time, but they are sometimes used to avoid confusion with other parallel lines, or to more clearly portray particular aspects of threads. There is a slight variation between ASA and CSA

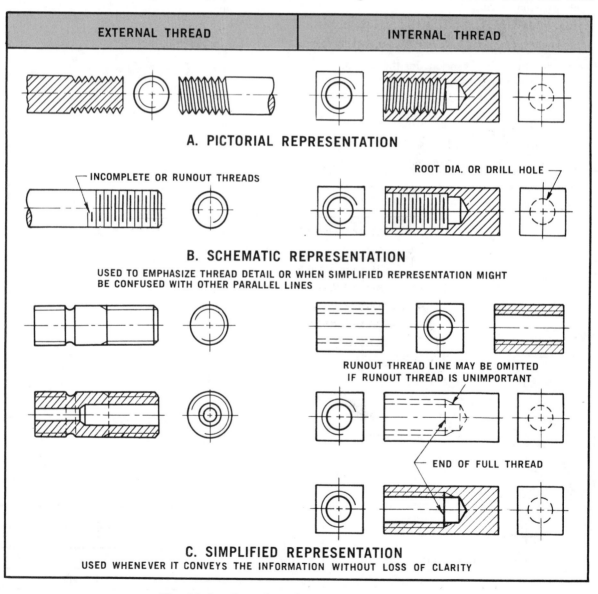

Fig. 11-1 Canadian Thread Representation.

thread representation as illustrated in figures 11-1 and 11-2.

Thread Standards

With the progress and growth of industry, a need has grown for uniform, interchangeable threaded fasteners. For convenience, several series of diameter-pitch combinations have been standardized. These series are Coarse, Fine, Extra Fine, and the Unified pitch series; that is, 8 thread, 12 thread, and 16 thread.

At present only the Coarse and Fine thread series from .250 inch up have been designated as UNC or UNF.

Thread Classes and Their Application

The fit of a screw thread is the amount of clearance between the screw and the nut when they are assembled together. In order to provide for various grades of fit, three classes of external threads (Classes 1A, 2A, and 3A) and three classes of internal threads (Classes 1B,

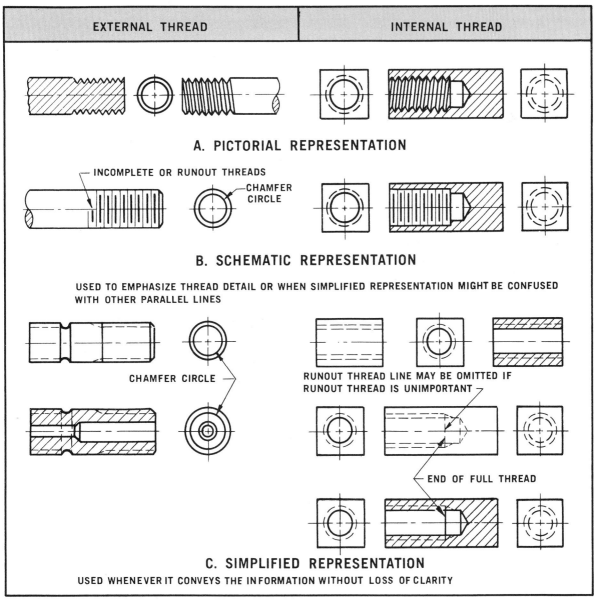

Fig. 11-2 American Thread Representation.

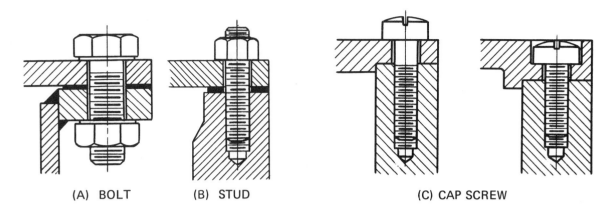

(A) BOLT   (B) STUD   (C) CAP SCREW

Fig. 11-3   Threaded Fasteners.

2B, and 3B) are provided in the Unified thread standard. These classes differ from each other in the amount of allowances and tolerances.

Right- and Left-Handed Threads

Unless designated otherwise, threads are assumed to be right hand. A cap screw being threaded into a tapped hole would be turned in a right-hand (clockwise) direction. For some special applications, such as turnbuckles, left-hand threads are required. When such a thread is necessary, the letters **LH** are added after the thread designation.

Thread Specifications

Information given for threads, whether external or internal, is expressed in this order: diameter (nominal or major diameter), number of threads per inch, thread form and series, and class of fit. The length of threads may be added to the note, or it may be shown on the drawing as a dimension. Figure 11-5 illustrates typical thread specifications.

KNURLING

Knurling is the roughing of a surface, as on the round head of a thumbscrew, to permit

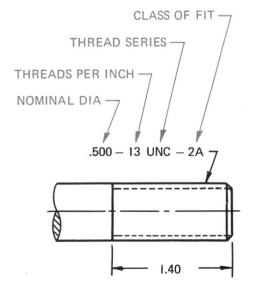

CLASS OF FIT
THREAD SERIES
THREADS PER INCH
NOMINAL DIA

.500 - 13 UNC - 2A

1.40

Fig. 11-4   Thread Designation.

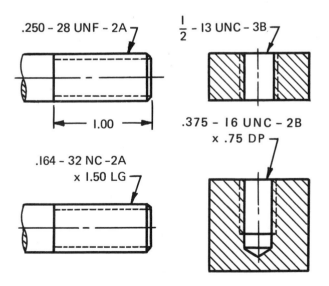

.250 - 28 UNF - 2A

1.00

.164 - 32 NC - 2A
x 1.50 LG

$\frac{1}{2}$ - 13 UNC - 3B

.375 - 16 UNC - 2B
x .75 DP

Fig. 11-5   Thread Specifications.

a better grip. Knurling is shown as either straight or diamond patterns. The pitch of the knurl may be specified. It is unnecessary to hatch in the whole area to be knurled if enough is shown so that the pattern can be clearly seen.

REFERENCES AND SOURCE MATERIALS

1. Canadian Standard Association, *Bulletin* 78.1 (1967).

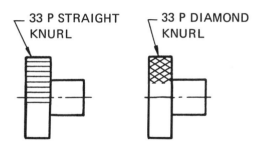

Fig. 11-6   Knurls.

QUESTIONS

1. Calculate distances (A) to (J)

2. How many threaded holes or shafts are shown?

3. How many chamfers are shown?

4. How many necks are shown?

5. What type of thread symbols are shown on this drawing?

Refer to Part 1

6. What are the limit sizes for the .64 dia. dimension?

7. What operation provides better gripping when turn-the clamping nut?

8. What is the size of the tap used?

9. Is the tap right hand or left hand?

10. How deep does the tap go?

Refer to Part 2

11. What is the length of thread?

12. What class of thread fit is required?

13. What heat treatment does the part undergo?

14. How many threads per inch are there?

Refer to Part 3

15. What does $\overset{F}{\vee}$ indicate?

16. What size thread is cut on the outside of the piece?

17. What is the clearance between the last thread and the flange?

18. What is the tolerance on the center-to-center distance between the tapped holes?

19. What is the smallest diameter to which the hole through the stuffing box can be made?

20. What are the limits for the 2.000 dia. dimension?

NOTE: EXCEPT WHERE NOTED —
TOLERANCE ON TWO DECIMAL
DIMENSIONS ± .02
TOLERANCE ON THREE DECIMAL
DIMENSIONS ± .005
TOLERANCE ON ANGLES ± 30'

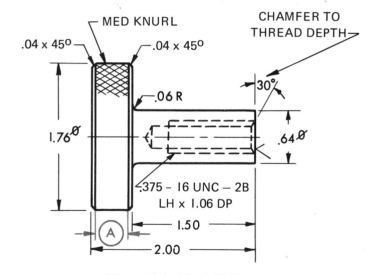

PT 1   CLAMPING NUT
MATL CRS, 16 REQD

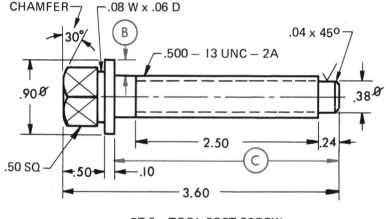

PT 2   TOOL POST SCREW
MATERIAL CRS CASE HARDEN, 8 REQD

58

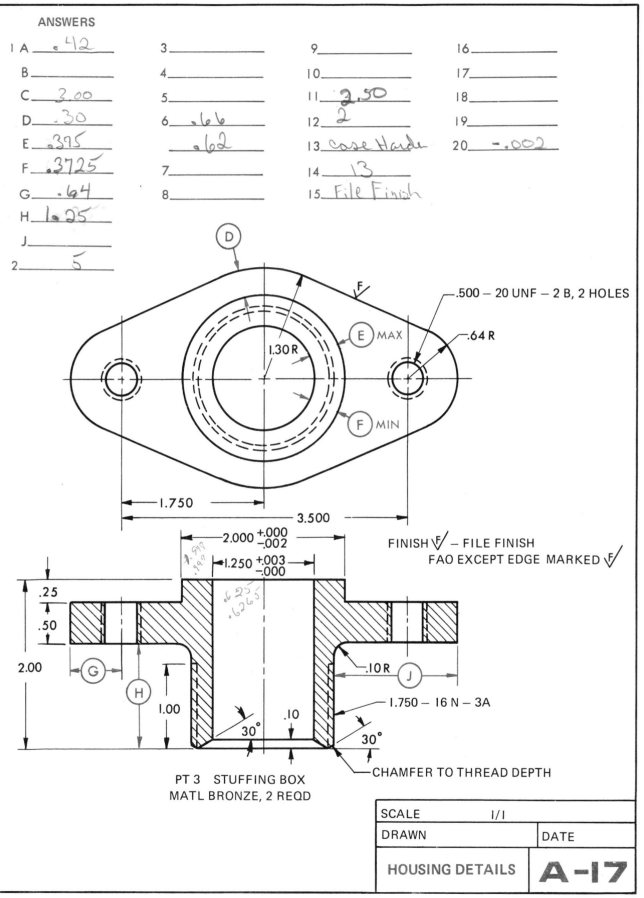

ANSWERS

1 A   .42
  B
  C   3.00
  D   .30
  E   .375
  F   .3725
  G   .64
  H   1.25
  J
2   5

3
4
5
6   .66
    .62
7
8

9
10
11   2.50
12   2
13   Case Harde
14   13
15   File Finish

16
17
18
19
20   -.002

.500 – 20 UNF – 2 B, 2 HOLES

.64 R

(D)

(E) MAX

(F) MIN

F

1.30 R

1.750

3.500

FINISH ⌵F – FILE FINISH
FAO EXCEPT EDGE MARKED ⌵F

2.000 +.000 −.002

1.250 +.003 −.000

.25

.50

2.00

(G)

(H)

1.00

30°

.10

.10 R

(J)

1.750 – 16 N – 3A

30°

CHAMFER TO THREAD DEPTH

PT 3   STUFFING BOX
MATL BRONZE, 2 REQD

SCALE	1/1	
DRAWN		DATE
HOUSING DETAILS		A-17

UNIT
12

REVOLVED AND REMOVED SECTIONS

Revolved and removed sections are used to show the cross-sectional shape of ribs, spokes, or arms when the shape is not obvious in the regular views. End views are often not needed when a revolved section is used.

REVOLVED SECTIONS

For a revolved section a centerline is drawn through the shape on the plane to be described, the part is imagined to be rotated 90 degrees, and the view that would be seen when rotated is superimposed on the view. If the revolved section does not interfere with the view on which it is revolved, then the view is not broken unless it would provide for clearer dimensioning. When the revolved sec-

tion interferes or passes through lines on the view on which it is revolved, then the general practice is to break the view. Often the break

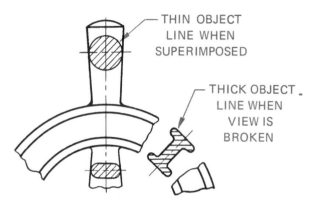

Fig. 12-1   Revolved Section. (Extracted from the American Standard Drafting Manual, Line Conventions, Sectioning and Lettering, ASA Y14-2-1957, with the permission of the Publisher, The American Society of Mechanical Engineers, 29 W. 39th St., New York, N.Y.)

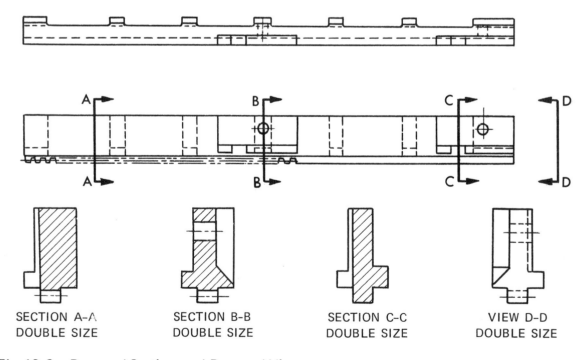

SECTION A-A
DOUBLE SIZE

SECTION B-B
DOUBLE SIZE

SECTION C-C
DOUBLE SIZE

VIEW D-D
DOUBLE SIZE

Fig. 12-2   Removed Sections and Removed View.   (Extracted from American Standard Drafting Manual, Line Conventions, Sectioning and Lettering, ASA Y14-2-1957, with permission of the Publisher, The American Society of Mechanical Engineers, 29 W. 39th St., New York, N.Y.)

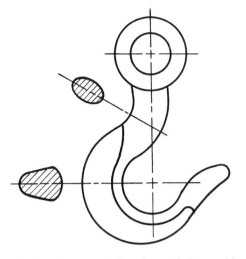

Fig. 12-3   Removed Section of Crane Hook.
(Courtesy CSA B 78.1 – 1967)

is used to shorten the length of the object. When superimposed on the view, the outline of the revolved section is a thin continuous line.

REMOVED SECTIONS

The removed section differs from the revolved section in that the section is removed to an open area on the drawing instead of being drawn right on the view. Frequently, the removed section is drawn to an enlarged scale for clarification and easier dimensioning.

Removed sections of symmetrical parts are placed on the extension of the centerline where possible.

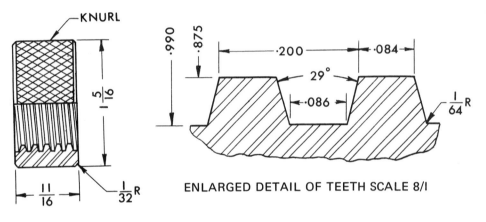

ENLARGED DETAIL OF TEETH SCALE 8/1

Fig. 12-4   Removed Section of Nut.

QUESTIONS

1. Calculate distances (A) to (K) .

2. How many surfaces require finishing?

3. What tolerance is permitted on three-decimal dimensions?

4. What type of section view is used on part 1?

5. What type of section view is used on part 2?

6. How many holes are there?

Refer to Part 1

7. What is the maximum center-to-center distance between the holes?

8. What surface finish is required?

9. Express the largest hole in plus or minus dimensioning.

10. What is the maximum permissible wall thickness at the larger hole?

11. What would be the minimum permissible wall thickness at the smaller hole?

Refer to Part 2

12. Name two machines that could produce the type of finish required for the slot.

13. What is the nominal size of machine screw used in the counterbored holes?

14. What type of machine screw would be used?

15. What is the distance between the center of the counterbored holes and the center of the slot?

NOTE: UNLESS OTHERWISE
        SPECIFIED —
ROUNDS AND FILLETS .10 R
TOLERANCE ON TWO-DECIMAL
    DIMENSIONS ± .02
TOLERANCE OF THREE-DECIMAL
    DIMENSIONS ± .005
FINISH ∨32∕

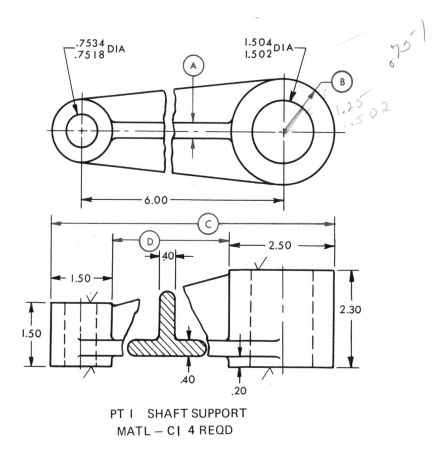

PT 1   SHAFT SUPPORT
MATL — CI 4 REQD

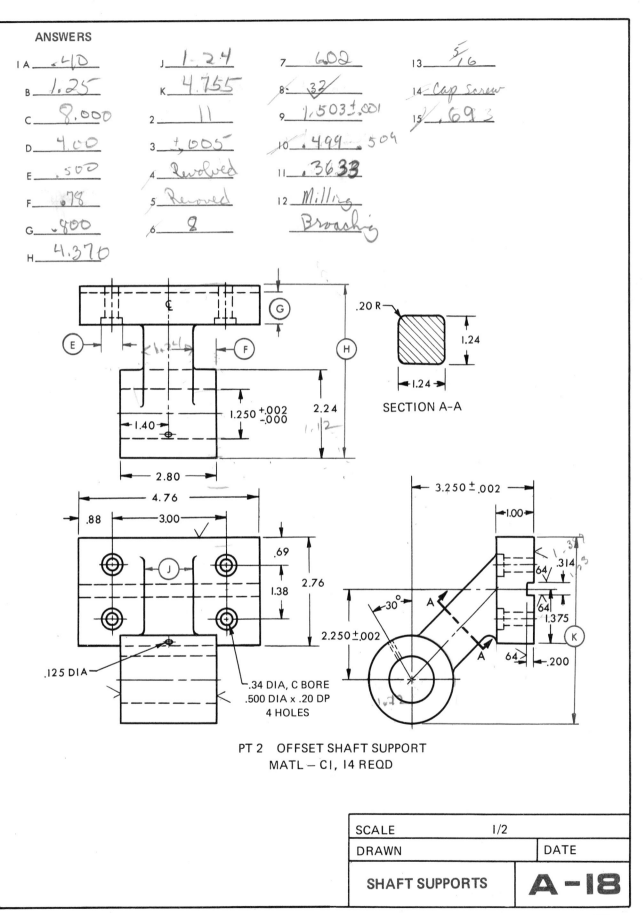

SECTION A-A

PT 2 OFFSET SHAFT SUPPORT

MATL – C1, 14 REQD

SCALE		1/2	
DRAWN			DATE
SHAFT SUPPORTS			A-18

63

UNIT 13

KEYS

A key is a piece of steel lying partly in a groove in the shaft, called a keyseat, and extending into another groove, called a keyway, in the hub. It is used to secure gears, pulleys, cranks, handles, and similar machine parts to shafts, so that the motion of the part is transmitted to the shaft, or the motion of the shaft to the part, without slippage. The key may also act in a safety capacity: its size is calculated so that, when overloading takes place, the key will shear or break before the part or shaft breaks.

There are many kinds of keys. The most common types are shown in figure 13-1. Flat and square keys are widely used in industry.

The Woodruff key is semicircular in shape and fits into a semicircular keyseat in the shaft and a rectangular keyway in the hub. Woodruff keys are identified by number. The key numbers indicate the nominal dimensions of the key. The last two digits of the number give the normal diameter in eighths of an inch, and the digits preceding the last two give the nominal width in thirty-seconds of an inch. For example, a No. 1210 Woodruff key

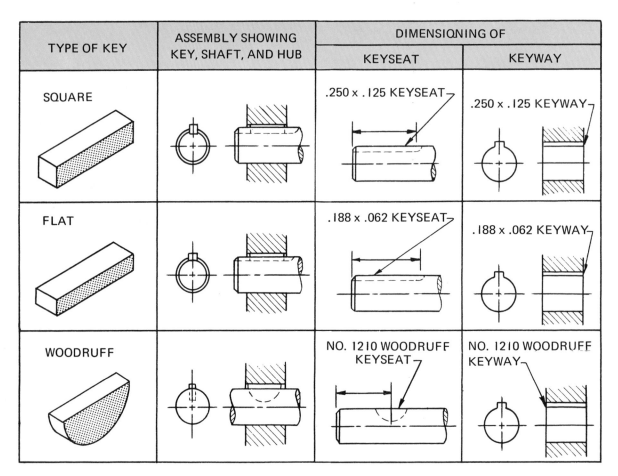

Fig. 13-1   Common Keys.

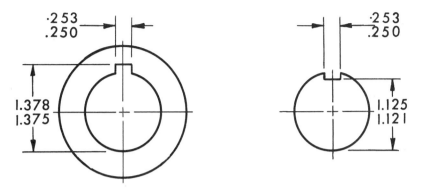

Fig. 13-2 Dimensioning Keyways and Keyseats for Interchangeable Assembly.

indicates a key 12/32 in. by 10/8 in., or a 3/8 in. by 1-1/4 in. key.

Dimensioning of Keyways and Keyseats

All dimensions of keyways and keyseats for square and flat keys, with the exception of the length of the flat portion of the keyseat which is given by a direct dimension on the drawing, are shown on the drawing by a note specifying first the width and then the depth. This type of dimensioning is the standard method used for unit production where the machinist is expected to fit the key into the keyway and keyseat.

For interchangeable assembly and mass production purposes, keyway and keyseat dimensions are given in limit dimensions to assure proper fits and are located from the opposite side of the hole or shaft.

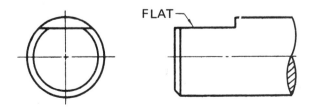

Fig. 13-3 Flats on Shafts.

FLATS

A flat is a slight depression usually cut on a shaft to serve as a surface on which the end of a setscrew can rest (when holding an object in place).

BOSSES AND PADS

A boss is a cylindrical projection of relatively small size above the surface of an object A pad is a slight projection, other than circular, above the surface of an object.

Bosses and pads should not be used in casting design unless absolutely necessary since the increased thickness in metal creates hot spots and causes open grain or draws in the casting.

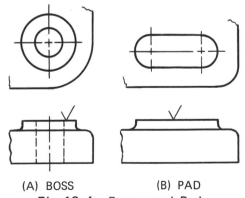

(A) BOSS   (B) PAD
Fig. 13-4 Bosses and Pads.

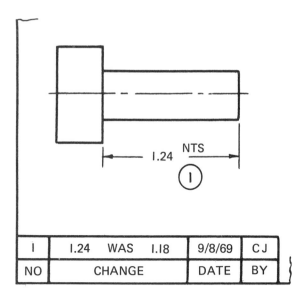

I	1.24 WAS 1.18	9/8/69	CJ
NO	CHANGE	DATE	BY

Fig. 13-5   A Typical Revision Note.

DRAWING REVISIONS

The term *drawing revision* refers to the changes made to a drawing after prints of the drawing are issued. Drawings are revised to improve design, improve manufacturing methods, reduce costs, correct errors and for many other reasons. A record is made of the change in a revision block showing the amount or kind of change, the date, and by whom it was made. The revision block is normally located adjacent to the borderline of the drawing.

Minor changes in size are frequently made without altering the original lines on a drawing. For this reason, a drawing should never be measured or scaled. A note or symbol indicating the dimension is not to scale is shown by the altered dimension. A typical drawing revision is shown in figure 13-5.

QUESTIONS

Refer to Part 1

1. What was the original length of the shaft?

2. What symbol is used to indicate that the $1\frac{7}{8}$ dimension is not to scale?

3. At how many places are threads being cut?

4. Specify for any left-hand threads the thread diameter and the number of threads per inch.

5. What class of fit is required for the threads?

6. What is the length of that portion of the shaft which has the (A) $\frac{7}{8}$ – 14 thread, (B) $1\frac{1}{4}$ – 12 thread, (C) 1 – 14 thread?

7. What distance is there between the last thread and the shoulder on the $\frac{7}{8}$ diameter shaft?

8. How many dimensions have been changed?

9. What type of section view is used?

10. What is the largest size to which the 1.250 diameter shaft can be turned?

11. What scale was used on this drawing?

Refer to Part 2

12. How many holes are there?

13. What are the overall length and width dimensions of the base?

14. How many surfaces are to be finished?

15. What is the diameter of the bosses on the base?

16. How wide is the pad on the upright shaft?

17. What is the depth of the keyway?

18. How far does the horizontal hole overlap the vertical hole? (Use maximum sizes.)

19. How much material was added when the change to the bosses was made?

20. What scale was used on this drawing?

21. What is the maximum permissible center-to-center distance of the two large holes?

ANSWERS

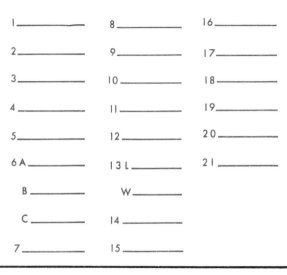

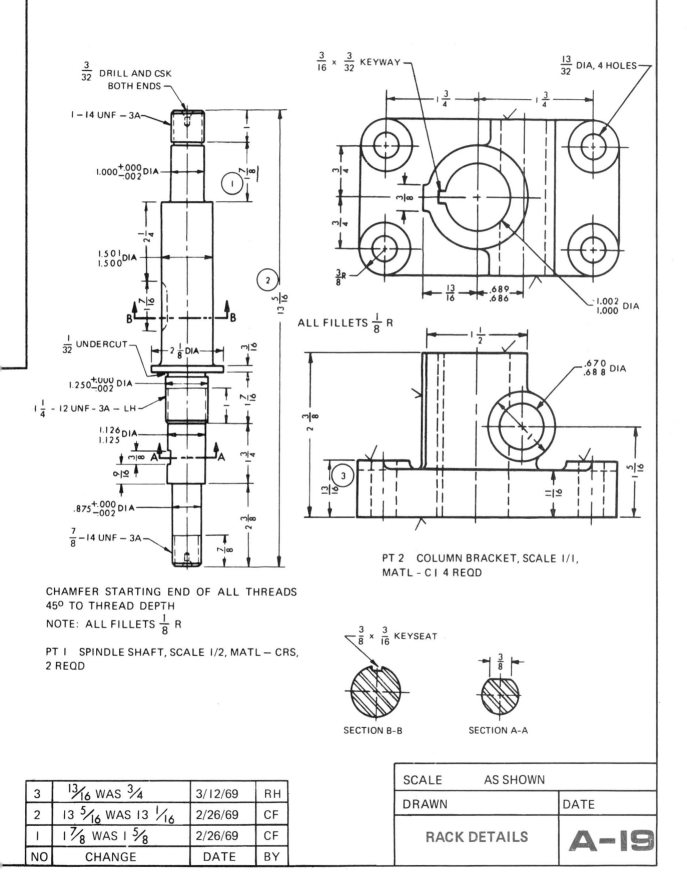

$\frac{3}{32}$ DRILL AND CSK BOTH ENDS

1 – 14 UNF – 3A

1.000 $^{+.000}_{-.002}$ DIA

1.501 / 1.500 DIA

$\frac{1}{32}$ UNDERCUT

1.250 $^{+.000}_{-.002}$ DIA

1 $\frac{1}{4}$ – 12 UNF – 3A – LH

1.126 / 1.125 DIA

.875 $^{+.000}_{-.002}$ DIA

$\frac{7}{8}$ – 14 UNF – 3A

CHAMFER STARTING END OF ALL THREADS 45° TO THREAD DEPTH

NOTE: ALL FILLETS $\frac{1}{8}$ R

PT 1   SPINDLE SHAFT, SCALE 1/2, MATL — CRS, 2 REQD

$\frac{3}{16} \times \frac{3}{32}$ KEYWAY

$\frac{13}{32}$ DIA, 4 HOLES

$\frac{3}{8}$ R

1.002 / 1.000 DIA

ALL FILLETS $\frac{1}{8}$ R

.670 / .688 DIA

PT 2   COLUMN BRACKET, SCALE 1/1, MATL – CI 4 REQD

$\frac{3}{8} \times \frac{3}{16}$ KEYSEAT

$\frac{3}{8}$

SECTION B-B   SECTION A-A

SCALE	AS SHOWN	
DRAWN		DATE
RACK DETAILS		**A-19**

3	$\frac{13}{16}$ WAS $\frac{3}{4}$	3/12/69	RH
2	13 $\frac{5}{16}$ WAS 13 $\frac{1}{16}$	2/26/69	CF
1	1 $\frac{7}{8}$ WAS 1 $\frac{5}{8}$	2/26/69	CF
NO	CHANGE	DATE	BY

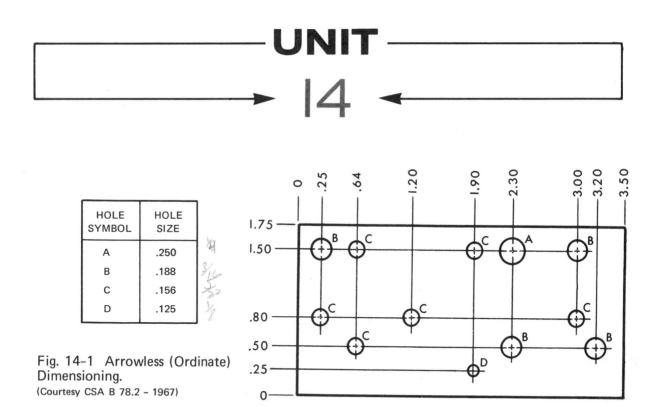

HOLE SYMBOL	HOLE SIZE
A	.250
B	.188
C	.156
D	.125

Fig. 14-1 Arrowless (Ordinate) Dimensioning.
(Courtesy CSA B 78.2 - 1967)

ARROWLESS DIMENSIONING [1]

To avoid having a large number of dimensions extending away from the part, arrowless or ordinate dimensioning may be used, as shown in figure 14-1. In this system, the "zero" lines represent the vertical and horizontal datum lines, and each of the dimensions shown without arrowheads indicates the distance from the zero line. There is never more than one zero line in each direction.

Arrowless dimensioning is particularly useful when such features are produced on a general-purpose machine, such as a jig borer, a tape-controlled drill, or a turret-type press.

HOLE SYMBOL	HOLE DIA.	LOCATION X→	Y↑
A₁	.250	2.30	1.50
B₁	.189	.25	1.50
B₂	.189	3.00	1.50
B₃	.189	2.30	.50
B₄	.189	3.20	.50
C₁	.159	.64	1.50
C₂	.159	1.90	1.50
C₃	.159	.25	.80
C₄	.159	1.20	.80
C₅	.159	3.00	.80
C₆	.159	.64	.50
D₁	.125	1.90	.25

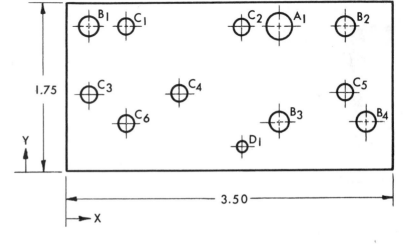

Fig. 14-2 Tabular Dimensions.
(Courtesy CSA B 78.2 — 1967)

TABULAR DIMENSIONING

When there are a very large number of holes or repetitive features, such as in a chassis or a printed circuit board, and where the multitude of centerlines would make a drawing difficult to read, tabular dimensioning is recommended. In this system each hole or feature is assigned a letter, or a letter and numeral subscript. The feature dimensions and the feature location along the X and Y axes are given in a table as shown in figure 14-2, page 68. The numbering and lettering of the features are normally from left to right and from top to bottom.

REFERENCES AND SOURCE MATERIALS

1. Canadian Standards Association, *Bulletin* 78.2 (1967).

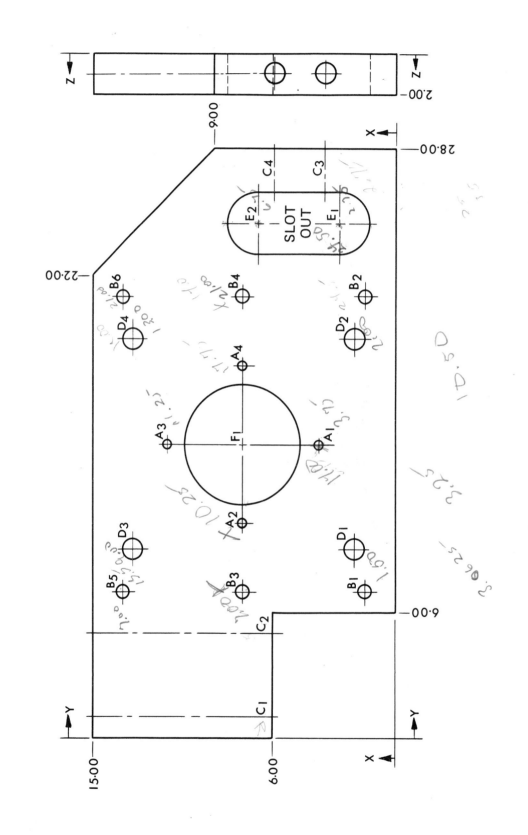

QUESTIONS

1. What is the length of the part?

2. What is the width of the part?

3. What is the length and width of the chamfer on the corner?

4. What is the width of hole **E**?

5. What is the length of hole **E**?

6. How much wood is left between hole **E** and the right-hand edge of the part?

7. What are the center distances between holes (A) B_5 and B_6; (B) D_3 and D_4; (C) B_1 and B_5; (D) D_2 and D_4; (E) A_1 and A_3; (F) C_1 and C_2; (G) B_3 and A_4?

8. How much wood is left between holes (A) A_1 and E_1; (B) A_2 and B_3; (C) C_1 and C_2; (D) B_3 and B_4?

ANSWERS

1 28.00

2 15.00

3 L 6.00 ~~458~~
   W 6.00

4 3.000

5 7.00

6 2.00

7 A 14.00
  B 10.00
  C 12.00
  D 11.00
  E 7.50
  F 4.00
  G 10.75

8 A 8.8125
  B 2.75
  C 3.88
  ~~13.375~~

$\dfrac{17}{18}$

HOLE SYMBOL	HOLE DIA.	LOCATION	
		X	Y
A_1	.375	3.75	14.00
A_2	.375	7.50	10.25
A_3	.375	11.25	14.00
A_4	.375	7.50	17.75
B_1	.625	1.50	7.00
B_2	.625	1.50	21.00
B_3	.625	7.50	7.00
B_4	.625	7.50	21.00
B_5	.625	13.50	7.00
B_6	.625	13.50	21.00

HOLE SYMBOL	HOLE DIA.	LOCATION		
		X	Y	Z
C_1	.812		1.00	1.00
C_2	.812		5.00	1.00
C_3	.812	3.50		1.00
C_4	.812	6.00		1.00
D_1	1.000	2.00	9.00	
D_2	1.000	2.00	19.00	
D_3	1.000	13.00	9.00	
D_4	1.000	13.00	19.00	
E_1	3.000	2.75	24.50	
E_2	3.000	6.75	24.50	
F_1	5.688	7.50	14.00	

QUANTITY	
MATERIAL	DRY MAPLE
SCALE	1/4
DRAWN	DATE
SUPPORT BRACKET	**A-20**

71

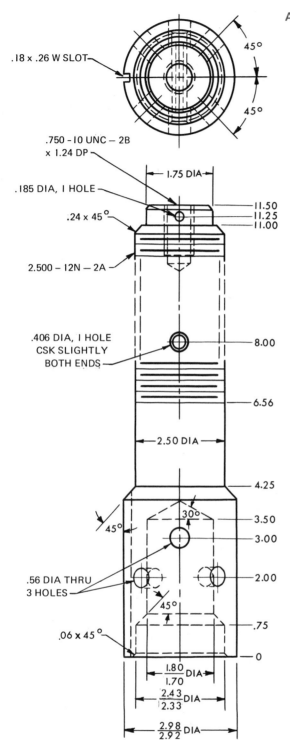

.18 x .26 W SLOT

.750 – 10 UNC – 2B
x 1.24 DP

.185 DIA, 1 HOLE

.24 x 45°

2.500 – 12N – 2A

.406 DIA, 1 HOLE
CSK SLIGHTLY
BOTH ENDS

.56 DIA THRU
3 HOLES

.06 x 45°

1.75 DIA

11.50
11.25
11.00

8.00

6.56

2.50 DIA

4.25

3.50
3.00

2.00

.75

0

30°

45°

45°

$\frac{1.80}{1.70}$ DIA

$\frac{2.43}{2.33}$ DIA

$\frac{2.98}{2.92}$ DIA

NOTE: INSIDE BORE AND THREADS TO BE
CLEAN AND BRIGHT, FAO ⌐250

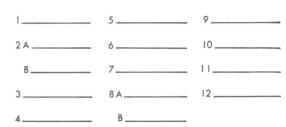

ANSWERS

1 _____	5 _____	9 _____
2 A _____	6 _____	10 _____
B _____	7 _____	11 _____
3 _____	8 A _____	12 _____
4 _____	B _____	

QUESTIONS

1. How long is the 2.50 dia. threaded section?

2. A. How deep is the .75 dia. tap?

   B. What type of threads are required?

3. How many .56 dia. holes are there?

4. What is the maximum thickness of the wall at the .56 dia. holes?

5. What class of fit is required for the tapped hole?

6. What is the distance between centerlines of the top .56 dia. hole and the .185 dia. hole?

7. What is the length of the 2.50 dia. unthreaded section?

8. A. How many threads per inch are there on the 2.50 dia. section?

   B. Approximately how many full threads are there?

9. What is the maximum overall dia. of the finished stud?

10. What tolerance is required on the largest bored hole at the bottom of the stud?

11. What is the quality of machining required?

12. What is the angle between the slot and the .56 dia. holes on the 2.00 dimension?

QUANTITY	3 REQD	
MATERIAL	COPPER	
SCALE	1/2	
DRAWN		DATE
TERMINAL STUD		A-21

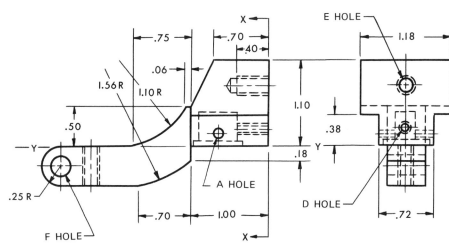

ANSWERS

1 _____
2 _____
3 _____
4 _____
5 _____
6 _____
7 _____
8 _____
9 _____
10 _____
11 _____
12 _____
13 _____
14 _____
15 _____

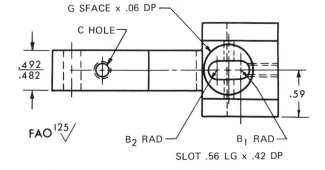

G SFACE x .06 DP
C HOLE
.492
.482
FAO ^{125} ⌄
B₂ RAD
B₁ RAD
SLOT .56 LG x .42 DP
.59

HOLE	HOLE SIZE	DISTANCE FROM	
		X–X	Y–Y
A	.125 DIA.	.66	.20
B₁	.12 RAD.	.50	
B₂	.12 RAD.	.82	
C	.190-24 UNC, 2B	2.30	
D	.164-32 UNC, 2B		.22
E	.250-20 UNC, 2B		.76
F	.189/.188 DIA.	2.72	.25
G	.656 DIA	.66	

QUESTIONS

1. What is the overall length?

2. What is the overall height?

3. What is the distance from **X–X** to the center of hole **F**?

4. What is the distance from **Y–Y** to **E** hole?

5. What is the horizontal distance between the centerlines of **A** and **F** holes?

6. What is the horizontal distance between the centerlines of **C** hole and **B₁** radius?

7. How many full threads does **E** hole have?

8. How deep is the spotface?

9. What is the tolerance on **F** hole?

10. What class of fit is required for the tapped hole **C**?

11. What does **UNC** mean?

12. What is the tolerance on the thickness of the material at **F** hole?

13. What is the length of the **C** hole?

14. What is the distance from line **X–X** to the termination of 1.10 radius?

15. What does **FAO** mean?

QUANTITY	8
MATERIAL	ALUMINUM
SCALE	1/1
DRAWN	DATE
CONTACTOR	**A-22**

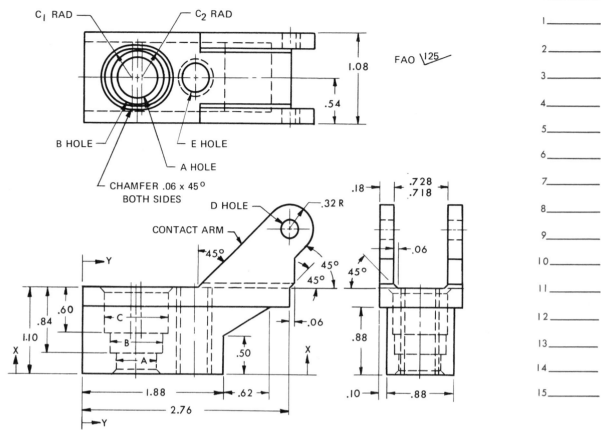

ANSWERS

1 _____
2 _____
3 _____
4 _____
5 _____
6 _____
7 _____
8 _____
9 _____
10 _____
11 _____
12 _____
13 _____
14 _____
15 _____

QUESTIONS

1. What is the overall length?

2. What is the overall height?

3. Give the chamfer angle.

4. What is the distance from **Y-Y** to **E** hole?

5. What is the distance from **X-X** to **D** hole?

6. How many complete threads has the tapped hole?

7. What class of fit is the tapped hole?

8. What surface finish is required?

9. How deep is the **C** counterbore?

10. How long is the **C** counterbore?

11. Give the distance between the ℄ of **C₁** and **C₂** counterbores.

12. What is the thickness of the contact arm at **D** hole?

13. What is the tolerance on the distance between the contact arms?

14. What is the nominal size of **D** hole?

15. What is the center distance between holes **A** and **E** ?

HOLE SYMBOL	HOLE SIZE	LOCATION X-X	LOCATION Y-Y
A	.531 DIA		.70
B	.672 DIA		.70
C_1	.359 RAD		.63
C_2	.359 RAD		.77
D	.250/.252 DIA	1.875	2.76
E	.50 – 13 UNC – 2B		1.48

QUANTITY		11
MATERIAL		MANGANESE BRONZE
SCALE		1/1
DRAWN		DATE

CONTACT ARM

A-23

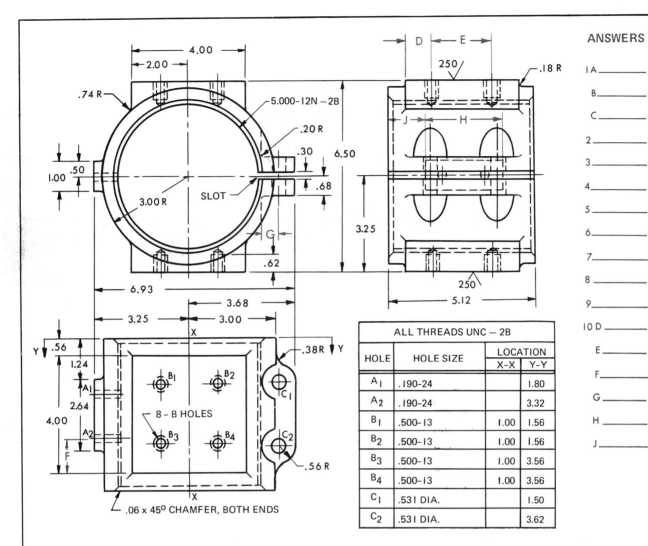

ANSWERS

I A_____
B_____
C_____
2_____
3_____
4_____
5_____
6_____
7_____
8_____
9_____
10 D_____
E_____
F_____
G_____
H_____
J_____

ALL THREADS UNC – 2B			
HOLE	HOLE SIZE	LOCATION	
		X–X	Y–Y
A₁	.190–24		1.80
A₂	.190–24		3.32
B₁	.500–13	1.00	1.56
B₂	.500–13	1.00	1.56
B₃	.500–13	1.00	3.56
B₄	.500–13	1.00	3.56
C₁	.531 DIA.		1.50
C₂	.531 DIA.		3.62

QUESTIONS

I. What is the center-to-center distance between the
   following?

   A. the A_1 and A_2 holes

   B. the B_1 and B_2 holes

   C. the C_1 and C_2 holes

2. What finish is required on the 4.00 x 4.00 face?

3. What is the width of the slot?

4. What series of threads are used on the **A** and **B**
   holes?

5. What threading is required in the 5-inch hole?

6. How deep are the **B** holes?

7. What size bolts would be used in the **C** holes?

8. How many chamfers are called for?

9. How many threaded holes are there?

10. Calculate distances **D, E, F, G, H, J.**

QUANTITY	24	
MATERIAL	COPPER	
SCALE	1/4	
DRAWN		DATE
TERMINAL BLOCK		A-24

UNIT

15

DETAILING OF PIPES AND TUBES

In many instances simplified drafting methods may be used in the detailing of pipes and tubes. The draftsman must decide which is the most efficient and timesaving method to use, and factors such as production methods and machines and the ability of the craftsman in the shop to interpret his drawings will influence his choice. Figure 15–1 gives two simplified methods for detailing the same tube.

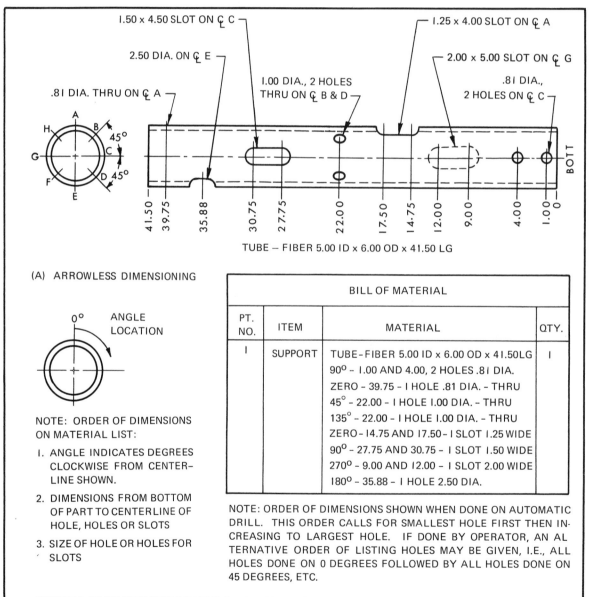

Fig. 15-1 Arrowless and Tabular Dimensioning for Pipes and Tubes.

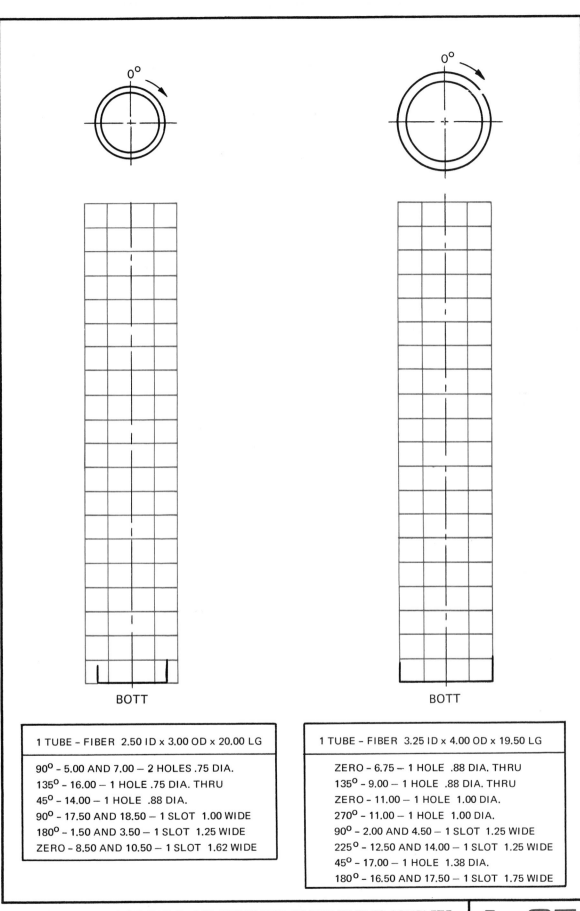

BOTT

BOTT

1 TUBE – FIBER 2.50 ID x 3.00 OD x 20.00 LG

90° – 5.00 AND 7.00 – 2 HOLES .75 DIA.
135° – 16.00 – 1 HOLE .75 DIA. THRU
45° – 14.00 – 1 HOLE .88 DIA.
90° – 17.50 AND 18.50 – 1 SLOT 1.00 WIDE
180° – 1.50 AND 3.50 – 1 SLOT 1.25 WIDE
ZERO – 8.50 AND 10.50 – 1 SLOT 1.62 WIDE

1 TUBE – FIBER 3.25 ID x 4.00 OD x 19.50 LG

ZERO – 6.75 – 1 HOLE .88 DIA. THRU
135° – 9.00 – 1 HOLE .88 DIA. THRU
ZERO – 11.00 – 1 HOLE 1.00 DIA.
270° – 11.00 – 1 HOLE 1.00 DIA.
90° – 2.00 AND 4.50 – 1 SLOT 1.25 WIDE
225° – 12.50 AND 14.00 – 1 SLOT 1.25 WIDE
45° – 17.00 – 1 HOLE 1.38 DIA.
180° – 16.50 AND 17.50 – 1 SLOT 1.75 WIDE

MAKE DRAWINGS OF THE TUBES DESCRIBED IN THE TABLES COMPLETE
WITH ARROWLESS DIMENSIONING. SCALE ¼ SIZE (ONE SQUARE = 1 INCH.)

A-25

ASSIGNMENT: IN THE SPACES PROVIDED, DETAIL THE TUBES SHOWN BELOW IN TABULAR
DIMENSIONING FORM, LISTING THE DETAILS ACCORDING TO THE COLUMN HEADINGS.
NOTE – ALL HOLES ON ONE SIDE ONLY UNLESS SPECIFIED BY THE WORD "THRU".

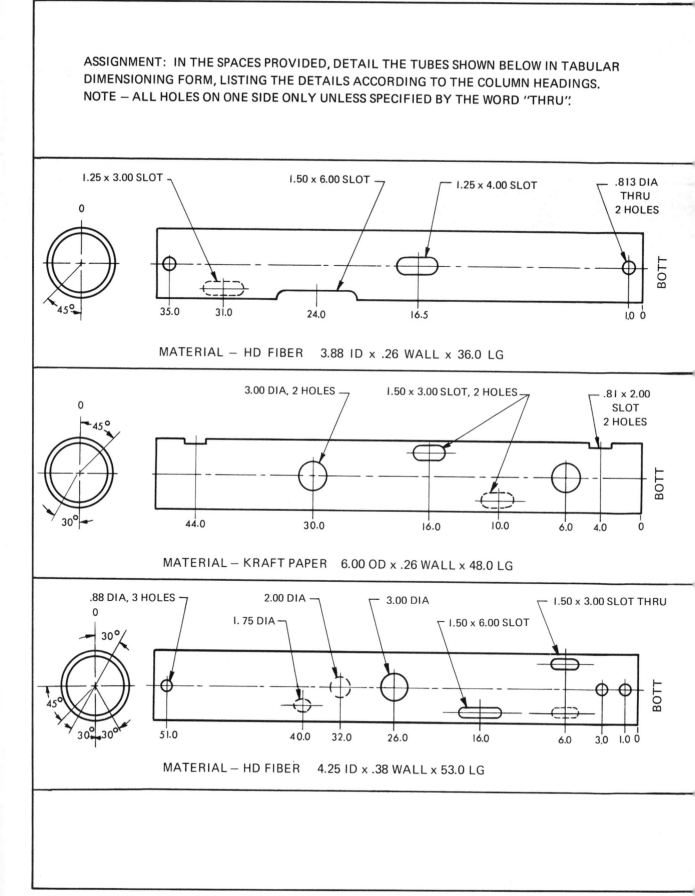

MATERIAL – HD FIBER   3.88 ID x .26 WALL x 36.0 LG

MATERIAL – KRAFT PAPER   6.00 OD x .26 WALL x 48.0 LG

MATERIAL – HD FIBER   4.25 ID x .38 WALL x 53.0 LG

COLUMN I	COLUMN 2
DETAIL THE TUBES BY CALLING FOR THE SMALLEST HOLE FIRST, THEN INCREASING TO THE LARGEST HOLE. 0	DETAIL THE TUBES BY TUBE ROTATION, I.E., ALL HOLES ON 0º, FOLLOWED BY ALL HOLES ON 45º, ETC. 0

SCALE	NONE	
DRAWN		DATE
TUBE SUPPORTS		A-26

STEEL SPECIFICATIONS [1]

Carbon steels are the workhorses of product design. They account for over 90 percent of total steel production. More carbon steels are used in product manufacturing than all other metals combined. Far more research is going into carbon steel metallurgy and manufacturing technology than into all other steel mill products.

The specifications covering the composition of metals have been issued by various classification bodies. They serve as a selection guide and provide a means for the buyer to conveniently specify certain known and recognized requirements. The main classification bodies are:

AISI — American Iron and Steel Institute
CSA — Canadian Standards Association
SAE — Society of Automotive Engineers

SAE and AISI — Systems of Steel Identification [2]

The specifications for steel bar are based on a code that indicates the composition of each type of steel covered. They include both plain carbon and alloy steels. The code is a 4-number system. Each figure in the number has the following specific function: the first or left-hand figure indicates the major class of steel, and the second figure represents a subdivision of the major class. For example, the series having *one* as the left-hand figure covers the carbon steels. The second figure breaks this class up into normal low-sulfur steels, the high-sulfur free machining grades and another grade having higher than normal manganese.

Class 1 —

Carbon Steels 1xxx

Basic open hearth and acid Bessemer carbon steels, nonsulfurized and nonphosphorized 10xx

Basic open hearth and acid Bessemer carbon steels, sulfurized but not phosphorized 11xx

Basic open hearth carbon steels phosphorized 12xx

Originally, this second figure indicated the percentage of the major alloying element present, and this is true of many of the alloy steels. However, this had to be varied in order to care for all the steels that are available.

The third and fourth figures indicate carbon content in hundredths of a percent; thus the figure xx15 indicates .15 of 1% carbon.

Example — SAE 2335 is a nickel steel containing 3.5% nickel, and .35 of 1% carbon.

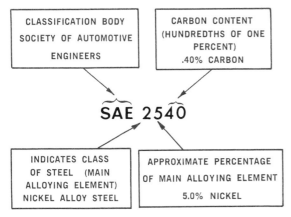

CLASSIFICATION BODY SOCIETY OF AUTOMOTIVE ENGINEERS	CARBON CONTENT (HUNDREDTHS OF ONE PERCENT) .40% CARBON

SAE 2540

INDICATES CLASS OF STEEL (MAIN ALLOYING ELEMENT) NICKEL ALLOY STEEL	APPROXIMATE PERCENTAGE OF MAIN ALLOYING ELEMENT 5.0% NICKEL

Fig. 16-1   Steel Designation System.

REFERENCES AND SOURCE MATERIALS

1. C.N. Parker, "Carbon Steels" *Machine Design* 37, No. 21 (1965).

2. Canadian Welding Bureau.

TYPE OF STEEL	NUMBER SYMBOL	PRINCIPLE PROPERTIES	COMMON USES
CARBON STEELS			
—Plain Carbon	10XX		
—Low Carbon Steel (0.6% to 0.20% Carbon)	1006 to 1020	Toughness and Less Strength	Chains, Rivets, Shafts, Pressed Steel Products
—Medium Carbon Steel (0.20% to 0.50% Carbon)	1020 to 1050	Toughness and Strength	Gears, Axles, Machine Parts, Forgings, Bolts and Nuts
—High Carbon Steel (Over 0.50% Carbon)	1050 and over	Less Toughness and Greater Hardness	Saws, Drills, Knives, Razors, Finishing Tools, Music Wire
—Sulphurized (Free Cutting)	11XX	Improves Machinability	Threads, Splines, Machined Parts
—Phosphorized	12XX	Increases Strength and Hardness but Reduces Ductility	
—Manganese Steels	13XX	Improves Surface Finish	
NICKEL STEELS	2XXX	Toughness and Strength	Crankshafts, Connecting Rods, Axles
—3.50% Nickel	23XX		
—5.00% Nickel	25XX		
NICKEL - CHROMIUM STEELS	3XXX	Toughness and Strength	Gears, Chains, Studs, Screws, Shafts
—0.70% Nickel 0.70% Chromium	30XX		
—1.25% Nickel 0.60% Chromium	31XX		
—1.75% Nickel 1.00% Chromium	32XX		
—3.50% Nickel 1.50% Chromium	33XX		
MOLYBDENUM STEELS	40XX	High Strength	Axles, Forgings, Gears, Cams, Mechanism Parts
—Chromium - Molybdenum Steels	41XX		
—Nickel - Chromium Molybdenum Steels	43XX		
—1.65% Nickel 0.25% Molybdenum	46XX		
—3.25% Nickel 0.25% Molybdenum	48XX		
CHROMIUM STEELS	5XXX	Hardness, Great Strength and Toughness	Gears, Shafts, Bearings, Springs, Connecting Rods
—Low Chromium	50XX		
—Medium Chromium	51XX		
—Chromium High Carbon	52XX		
CHROMIUM VANADIUM STEELS	61XX	Hardness and Strength	Punches and Dies, Piston Rods, Gears, Axles
NICKEL - CHROMIUM - MOLYBDENUM STEELS	86XX	Rust Resistance, Hardness and Strength	Food Containers, Surgical Equipment
SILICON - MANGANESE STEELS	92XX	Springiness and Elasticity	Springs

Fig. 16–2 Designations, Uses, and Properties of Steel.

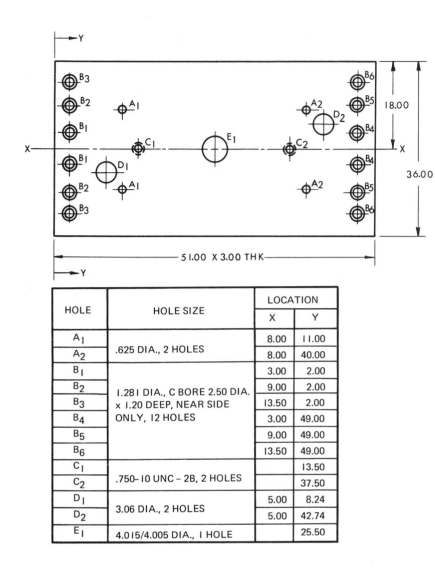

HOLE	HOLE SIZE	LOCATION	
		X	Y
A_1	.625 DIA., 2 HOLES	8.00	11.00
A_2		8.00	40.00
B_1	1.281 DIA., C BORE 2.50 DIA. x 1.20 DEEP, NEAR SIDE ONLY, 12 HOLES	3.00	2.00
B_2		9.00	2.00
B_3		13.50	2.00
B_4		3.00	49.00
B_5		9.00	49.00
B_6		13.50	49.00
C_1	.750–10 UNC – 2B, 2 HOLES		13.50
C_2			37.50
D_1	3.06 DIA., 2 HOLES	5.00	8.24
D_2		5.00	42.74
E_1	4.015/4.005 DIA., 1 HOLE		25.50

QUESTIONS

1. What is the thickness of the material?

2. What type of steel is specified?

3. What tells you that the drawing is not to scale?

4. How many counterbored holes are there?

5. What is the diameter of the C-bore?

6. How many 3.06 dia. holes are there?

7. What letter specifies the 3.06 dia. holes?

8. How many tapped holes of different sizes are specified?

9. What is the total number of tapped holes?

10. What class of fit is specified for the C holes?

11. What is the tolerance given for the E hole?

12. What is the high limit of the E hole?

13. What would be the maximum limit for 3.06 dia. hole if a $\pm$ .004 leeway is permitted?

14. Is the E hole on the center of the part?

15. How far apart are holes C_1 and C_2?

QUANTITY	4	
MATERIAL	SAE 1020	
SCALE	NONE	
DRAWN		DATE

CROSSBAR **A-27**

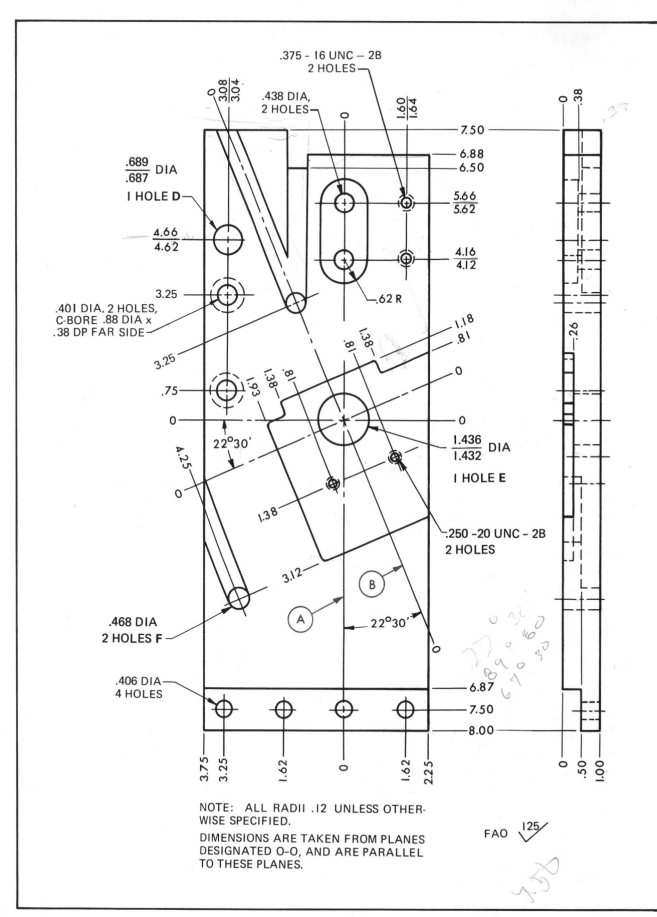

.375 - 16 UNC - 2B
2 HOLES

.438 DIA,
2 HOLES

$\frac{.689}{.687}$ DIA

I HOLE **D**

$\frac{4.66}{4.62}$

.401 DIA, 2 HOLES,
C-BORE .88 DIA x
.38 DP FAR SIDE

3.25

3.25

.75

0

22°30'

0

.468 DIA
2 HOLES **F**

.406 DIA
4 HOLES

3.08 / 3.04

1.60 / 1.64

7.50
6.88
6.50

$\frac{5.66}{5.62}$

$\frac{4.16}{4.12}$

.62 R

1.18
.81
1.38
.81

0

$\frac{1.436}{1.432}$ DIA

I HOLE **E**

.250 –20 UNC – 2B
2 HOLES

1.93
1.38
.81

1.38

4.25

0

3.12

(A)   (B)

22°30'

0

6.87
7.50
8.00

3.75 / 3.25   1.62   0   1.62 / 2.25

0 / .38

.26

0   .50 / 1.00

NOTE: ALL RADII .12 UNLESS OTHER-
WISE SPECIFIED.

DIMENSIONS ARE TAKEN FROM PLANES
DESIGNATED O–O, AND ARE PARALLEL
TO THESE PLANES.

FAO   125

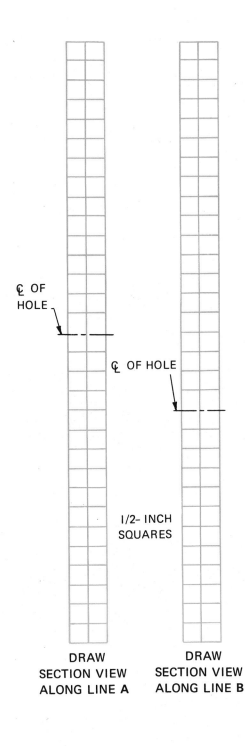

C OF
HOLE

C OF HOLE

1/2- INCH
SQUARES

DRAW
SECTION VIEW
ALONG LINE A

DRAW
SECTION VIEW
ALONG LINE B

QUESTIONS

1. What is the overall (A) width, (B) length, (C) thickness of the parts?
2. What is the width of the slots that are cut out from holes **F**?
3. What is their depth?
4. What class of machining is required?
5. What is the depth of the recess adjacent to the (A) **E** hole, (B) .438 dia. holes?
6. How many degrees are there between line **B** and the horizontal?
7. What class of fit is required on the tapped holes?
8. What tolerance is required on the **E** hole ?
9. What is its low limit on the **E** hole ?
10. What tolerance is required on the **D** hole?
11. What is its high limit on the **D** hole?
12. Determine the maximum vertical distance from **D** hole to the .406 dia. holes.
13. Determine the (A) maximum, (B) minimum, distance between .375 tapped holes.
14. Determine the maximum horizontal distance between the **D** hole and the .375 tapped holes.
15. What is the depth, width, and length of the cutout at the bottom of the part?
16. How deep is (A) the .438 dia. hole, (B) the .406 dia. hole?
17. How deep are the .250 tapped holes?
18. How many full threads have the .250 tapped holes?

ANSWERS

1 A __6.00__

B __15.50__

C __1.00__

2 __.468__

3 __.50__

4 __FAO 125__

5 A __.26__

B __.38__

6 __67°30'__

7 __2B__

8 __±.002__

9 __1.432__

10 __±.001__

11 __.689__

12 __12.16__

13 A __1.54__

B __1.46__

14 __4.72__

15 D __.50__

W __6.00__

L __1.13__

16 A __.62__

B __.50__

17 __.74__

18 __14.__

QUANTITY	7	
MATERIAL	SAE 40 20	
SCALE	1/2	
DRAWN		DATE

OIL CHUTE A-28

85

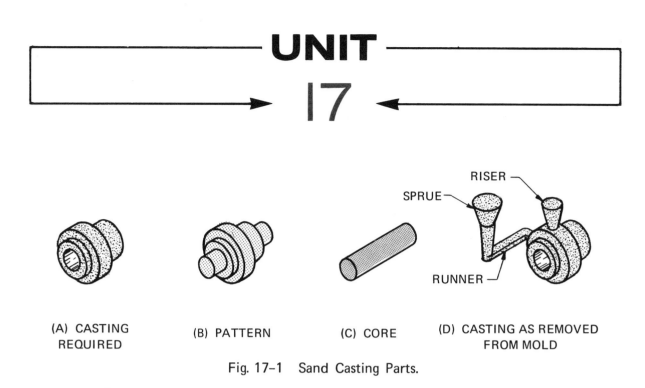

(A) CASTING REQUIRED (B) PATTERN (C) CORE (D) CASTING AS REMOVED FROM MOLD

Fig. 17-1 Sand Casting Parts.

CASTINGS [1]

Irregular or odd-shaped parts which are difficult to make from metal plate or bar stock may be cast to the desired shape. Casting processes for metals can be classified by either the type of mold or pattern or the pressure or force to fill the mold. Conventional sand, shell, and plaster molds utilize a durable pattern, but the mold is used only once. Permanent molds and die-casting dies are machined in metal or graphite sections and are employed for a large number of castings. Investment casting and the relatively new full-mold process involve both an expendable mold and pattern.

Sand Mold Casting

The most widely used casting process for metals uses a permanent pattern of metal or wood that shapes the mold cavity when loose molding material is compacted around the pattern. This material consists of a relatively fine sand plus a binder that serves as the refractory aggregate.

Figure 17-2 shows a typical sand mold, with the various provisions for pouring the molten metal and compensating for contraction of the solidifying metal, and a sand core for forming a cavity in the casting. Sand molds consist of two or more sections: bottom (drag), top (cope), and intermediate sections (checks) when required.

The sand is contained in flasks equipped with pins and lugs to insure the alignment of the cope and drag. Molten metal is poured into the sprue, and connecting runners provide flow channels for the metal to enter the mold cavity through gates. Riser cavities are located over the heavier sections of the casting.

The gating system, besides providing a way for the molten metal to enter the mold, functions as a venting system for the removal of gases from the mold and acts as a riser to furnish liquid metal to the casting during solidification.

In producing sand molds, a metal or wooden pattern must first be made. The pattern, normally made in two parts, is slightly larger in every dimension than the part to be cast to allow for shrinkage when the casting cools. This is known as shrinkage allowance, and the patternmaker allows for it by using a shrink rule for each of the cast metals. Since

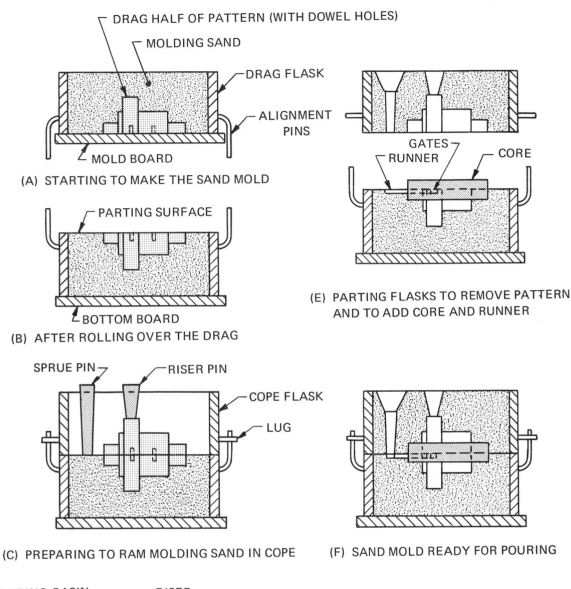

(A) STARTING TO MAKE THE SAND MOLD

(B) AFTER ROLLING OVER THE DRAG

(C) PREPARING TO RAM MOLDING SAND IN COPE

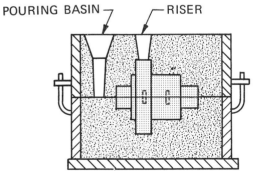

(D) REMOVING RISER AND GATE SPRUE PINS AND ADDING POURING BASIN

(E) PARTING FLASKS TO REMOVE PATTERN AND TO ADD CORE AND RUNNER

(F) SAND MOLD READY FOR POURING

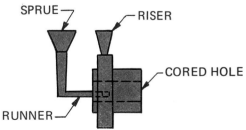

SPRUE, RISER, AND RUNNER TO BE REMOVED FROM CASTING.

(G) CASTING AS REMOVED FROM THE MOLD

Fig. 17-2   Sequence in Preparing a Sand Casting.

shrinkage and draft are taken care of by the patternmaker, they are of no concern to the draftsman.

Additional metal, known as machining or finish allowance, must be provided on the casting where a surface is to be finished. Depending on the material being cast, between .06 and .12 in. is usually allowed on small castings for each surface requiring finishing.

When casting a hole or recess in a casting, a core is often used. A core is a mixture of sand and a bonding agent that is baked and hardened to the desired shape of the cavity in the casting, plus an allowance to support the core in the sand mold. In addition to the shape of the casting desired, the pattern must be designed to produce areas in the mold cavity to locate and hold the core. The core must be solidly supported in the mold, permitting only that part of the core that corresponds to the shape of the cavity in the casting to project into the mold.

Preparation of Sand Molds. The drag portion of the flask is first prepared in an upside-down position. The drag half of the pattern is placed in position on the mold board, and the molding sand is then rammed or pressed into the drag flask. A bottom board is placed on the lower drag, the whole unit is rolled over, and the mold board is removed.

The cope half of the pattern is placed in the drag half and the cope portion of the flask is placed in position. Next the sprue pin and riser pin, tapered for easy removal, are located, and the molding sand is rammed into the cope flask. The sprue pin and riser pin are then removed. A pouring basin may or may not be formed at the top of the gate sprue.

Now the cope is lifted carefully from the drag, and the pattern is removed. The runner and gate, passageways for the molten metal to flow into the mold cavity, are formed in the drag sand, and the core is placed in position. The cope is then put back on the drag.

The molten metal is poured into the pouring basin and runs down the gate sprue to a runner and through the gate and into the mold cavity.

When the metal has hardened, the sand is broken and the casting removed. The excess metal, gates and risers, are removed and then remelted.

Full Mold Casting

The characteristic feature of the full mold process is the use of gasifiable patterns made of foamed plastic. These are not extracted from the mold, but are vaporized by the molten metal.

The full mold process is suitable for individual castings, and for small series of up to five castings. The advantages it offers are obvious: it is very economical, and it reduces the delivery time required for prototypes, articles urgently needed for repair jobs, and individual large machine parts.

REFERENCES AND SOURCE MATERIALS

1. J.F. Wallace. "Casting" *Machine Design*, 37, No. 21 (1965).

QUESTIONS

1. What line in the top view represents the same surface represented by line (A) ?

2. How many holes are there in the bracket?

3. How many of the holes are to be drilled?

4. What machining operation could be used to produce the largest hole?

5. Dimension (N) has been omitted. What should it be?

6. (A) How many ribs are there and (B) how thick are they?

7. Determine distances (B)(C)(E)(F)(G) (J) and (M)

8. What is the overall length of the part?

9. What is the overall height of the part?

10. What is the overall width of the part?

11. What is the name given to the operation at (D) ?

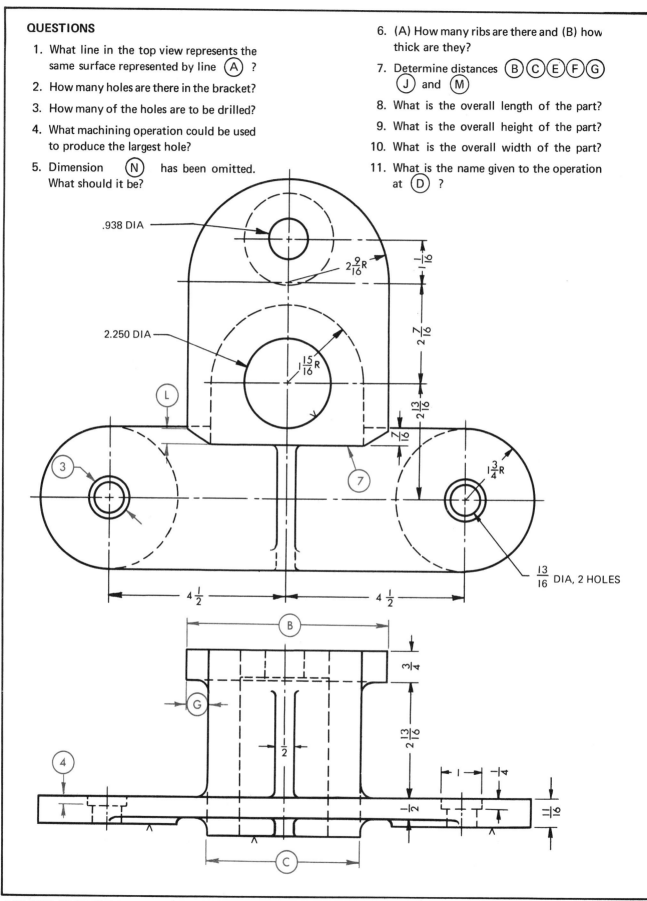

ADAPTED FROM "MECHANICAL DRAWING," HENRY FORD TRADE SCHOOL, DEARBORN, MICHIGAN

12. How many definite surfaces are to be finished?

13. What is the operation called when a hole is made larger, as at (3) and (4) ?

14. What is the diameter at (3) ?

15. Give depth of the spotface.

16. The draftsman neglected to put a dimension at (L) . What should it be?

17. Assuming that 1/8 in. extra has been allowed on the casting for all surfaces to be finished, what would distance (K) be on the rough casting?

18. Determine distance or dimensions for machining operations at (P) (Q) (R) (S) (T) (W) (Y) (Z)

ANSWERS

(E)	13		18	(U)
1	(F)	14		(V)
2	(G)	15		(W)
3	(J)	16		(X)
4	(M)	17		(Y)
5	8		18	(P) (Z)
6 A	9		(Q)	
B	10		(R)	
7 (B)	11		(S)	
(C)	12		(T)	

ROUNDS AND FILLETS $\frac{1}{8}$ R

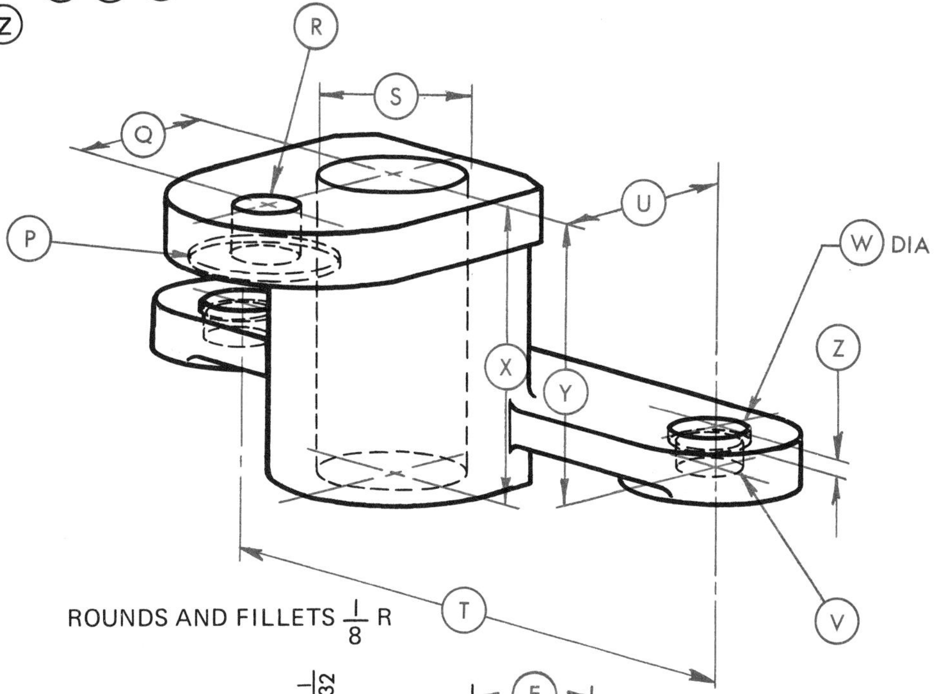

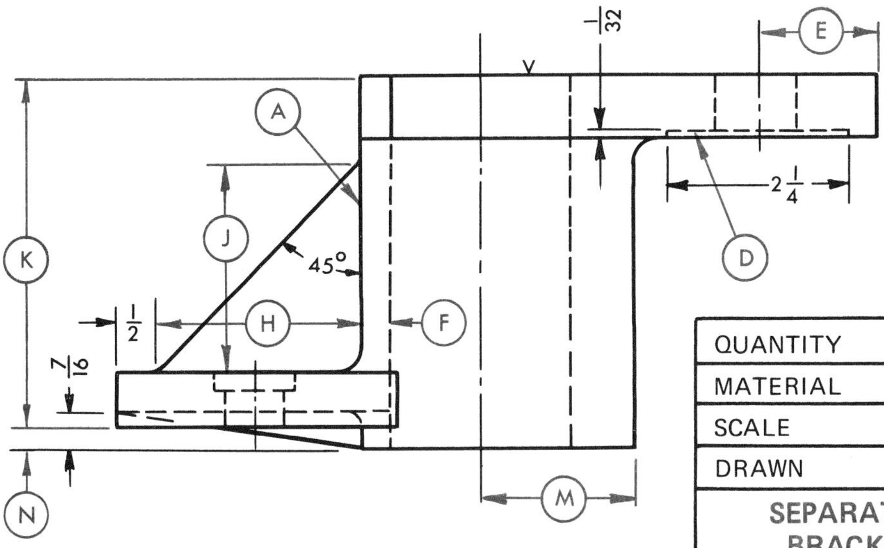

45°

$\frac{1}{32}$

$2\frac{1}{4}$

$\frac{1}{2}$

$\frac{7}{16}$

QUANTITY	24	
MATERIAL	CI	
SCALE	1/2	
DRAWN		DATE
SEPARATOR BRACKET		A-29

91

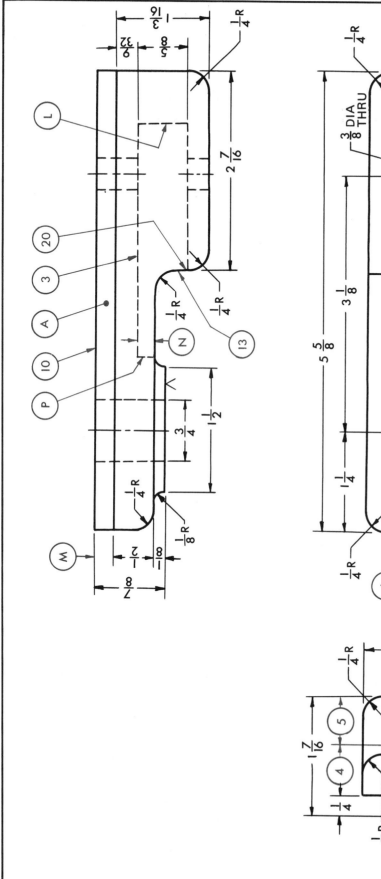

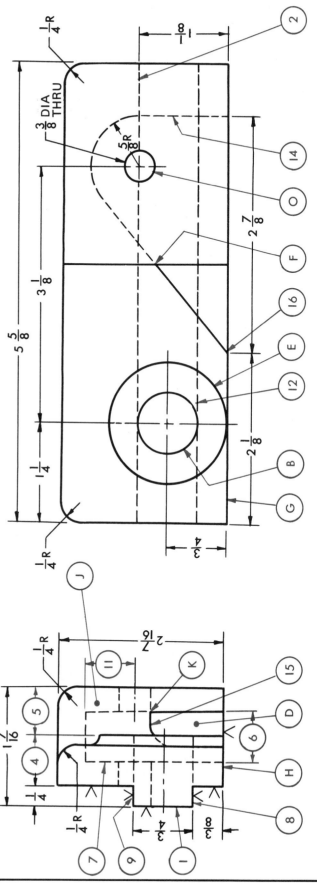

QUESTIONS

1. What line in the top view represents surface ①?

2. Locate surface Ⓐ in the left view and the front view.

3. Locate surface ⑧ in the front view.

4. How many surfaces are to be finished?

5. What line in the left view represents surface ③?

6. What is the center distance between holes Ⓑ and Ⓞ?

7. Determine distances at ④⑤⑥ and ⑪.

8. Locate surface Ⓙ in the top view.

9. What surface of the left view does line ⑭ represent?

10. What point in the front view represents line ⑮?

11. What is the thickness of boss Ⓔ?

12. Locate surface Ⓖ in the left view.

13. Locate point Ⓚ in the top view.

14. Locate surface Ⓓ in the top view.

15. Determine distance at Ⓜ Ⓝ.

16. What point or line in the top view is represented by point ⑯?

ANSWERS

1 _____

2 L.V. _____

  F.V. _____

3 _____

4 _____

5 _____

6 _____

7 ④ _____ ⑤ _____ ⑥ _____ and ⑪ _____

8 _____

9 _____

10 _____

11 _____

12 _____

13 _____

14 _____

15 Ⓜ _____ Ⓝ _____

16 _____

QUANTITY	5
MATERIAL	C I
SCALE	1/1
DRAWN	DATE

TRIP BOX

A-30

CASTING DESIGN

Simplicity of Molding From Flat Back Patterns

Simple shapes such as the one shown in figure 18-1 are very easy to mold. In this case the flat face of the pattern is at the parting line and lies perfectly flat on the molding board. In this position no molding sand sifts under the flat surface to interfere with the drawing of the pattern. The simplicity with which flat back patterns of this type may be drawn from the mold is illustrated.

Irregular or Odd-Shaped Castings

When a casting is to be made for an odd-shaped piece such as the offset bracket, figure 18-2, it is necessary to make the pattern for the bracket in one or more parts to facilitate the making of the mold. The difficulties encountered are due almost entirely to the removal of the pattern from the sand mold.

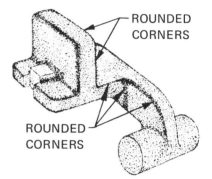

ROUNDED CORNERS

ROUNDED CORNERS

Fig. 18-2 Casting of Offset Drawing Shown in Drawing A-31.

The pattern for the bracket is made in two parts as shown in figure 18-3. The two adjacent flat surfaces of the divided pattern come together at the parting line.

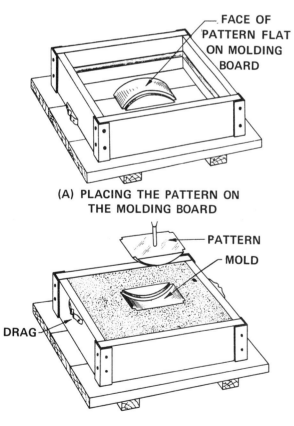

FACE OF PATTERN FLAT ON MOLDING BOARD

(A) PLACING THE PATTERN ON THE MOLDING BOARD

PATTERN

MOLD

DRAG

(B) DRAWING THE PATTERN

Fig. 18-1   Making a Mold of a Flatback Pattern.

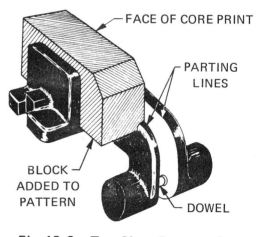

FACE OF CORE PRINT

PARTING LINES

BLOCK ADDED TO PATTERN

DOWEL

Fig. 18-3   Two-Piece Pattern of Offset Bracket.

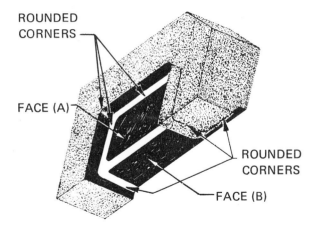

ROUNDED CORNERS

FACE (A)

ROUNDED CORNERS

FACE (B)

Fig. 18-4   Set Core for Offset Bracket.

Set Cores

In examining the illustration of the offset bracket, it will be noted that there are rounded corners. In order to make it possible to mold these rounded corners, a block must be added to the pattern for ease in removing from the mold.

This block, which becomes an integral part of the pattern, also acts as a core print for a set core, figure 18-4. It is made so that it conforms to the shape of the faces of the casting, including the rounded corners.

The face of the core print also forms the parting line for one side of the two-part pattern. When making the mold for the bracket casting, this face, which corresponds to the flat face of a flat back pattern, is laid on the mold-ing board with the drag of the flask in position. The sand is then rammed around the pattern.

When the drag is reversed, the cope, or upper part of the flask, is placed in position. The other half of the pattern is then joined with the first part, and the sand is rammed into the cope to flow around that part of the pattern which projects into it.

After the pattern is removed, the set core, which is formed in a core box and baked hard, is set in the impression in the mold made by the core print of the pattern. When poured into the mold, the molten metal fills the cavity made by the pattern and the faces of the set core as shown in figure 18-4 at **A** and **B** to form the casting.

Coping Down

The core print is made as part of the pattern to avoid removing molding sand in the drag which would correspond to the shape of the core print.

If this sand were dug out or *coped out,* as shown in figure 18-5, the remaining cavity would be again filled with molding sand when the cope was rammed. The sand would then hang below the parting line of the cope down into the drag.

When the mold is made by *coping down,* the hanging portion of the cope is supported by *soldiers* or *gaggers* imbedded within the

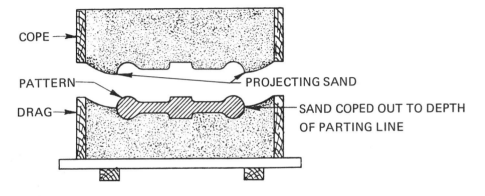

COPE

PATTERN

DRAG

PROJECTING SAND

SAND COPED OUT TO DEPTH OF PARTING LINE

Fig. 18-5   Coping Down.

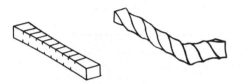

Fig. 18–6 Soldiers and Gaggers.

sand to hold the projecting part in position for subsequent operations.

Coping down requires skill and takes time. The *set core* principle is preferred at times to coping down in order to avoid such delay and to assure a more even parting line on the casting.

Split Patterns

Irregularly shaped patterns which cannot be drawn from the sand are sometimes split so that one half of the pattern may be rammed in the drag as a simple flat back pattern while the cope half, when placed in position on the drag half, forms the mold in the cope. Patterns of this type are called *split patterns* and do not require coping down to the parting line which would otherwise be necessary if the pattern were made solid.

The pattern for the casting shown in figure 18-7, when made without a print for a core, can be drawn from the sand only by

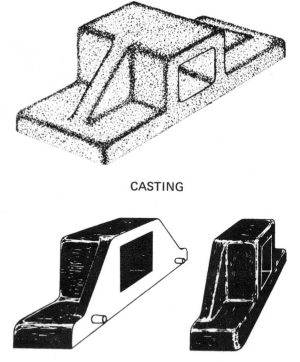

CASTING

SPLIT-PATTERN

Fig. 18–7 Application of Split Pattern.

splitting the pattern on the parting line as illustrated.

The drawing must be examined to determine how the pattern is to be constructed so that the parting line can be located in such a position to permit the halves of the pattern to be drawn from the sand without interference.

then measured and checked against the computed value of **R** as found by the following formula:

$$R = M - [D \ (1 + \cot \frac{\text{angle X}}{2})]$$

Note: Diameter **D** should be slightly less than distance **L**.

Example:

Given: Distance **M** = 3.000″
Diameter of rod **D** = .625″
Degrees in angle **X** = 55°

Then:

R = 3.000 - .625 (1 + 1.921)
Combining: R = 3.000 - 1.826
R = 1.174″

Male dovetails may also be measured in a similar manner by placing two accurate rods of known diameter against the sides and bottom of the dovetail. The distance **Q** over the rods is measured and then checked against the computed value of **Q** as found by the following formula:

$$Q = D \ (1 + \cot \frac{\text{angle X}}{2}) + S$$

Example:

Given: Distance **S** = 2.000″
Diameter of rod **D** = .625″
Degrees in angle **X** = 55°

Then: Q = .625 (1 + 1.921) + 2
Combining: Q = (.625 x 2.921) + 2
Q = 3.826″

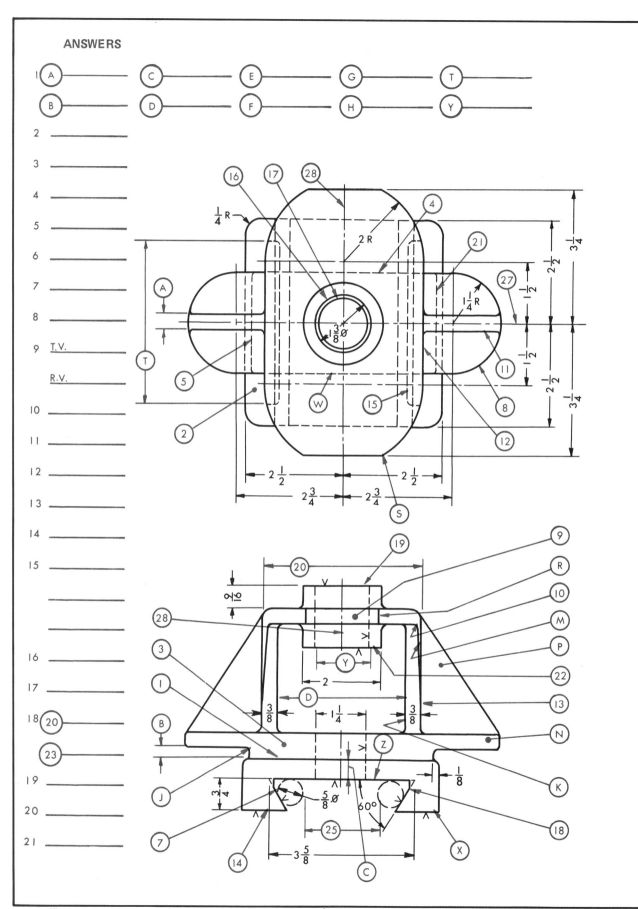

QUESTIONS

1. Determine distance at (A)(B)(C)(D)(E)(F)(G) (H)(T)(Y).
2. What surface in the top view does line (I) represent?
3. What line or surface in the top view does line (J) represent?
4. What line or surface in the top view does line (K) represent?
5. On what line would the pattern for the shuttle be split?
6. What line in the right view represents line (M) ?
7. What line in the top view represents surface (N) ?
8. What line or surface in the top view is represented by line (O) ?
9. What lines in the top view and right view represent surface (P) ?

10. What surface in the front view is represented by line (Q) ?
11. What line in the right view represents line (R) ?
12. What line in the right view represents corner (S) ?
13. What point in the front view shows point (U) ?
14. What surface in the front view represents line (W) ?
15. What surfaces are finished?
16. What line in the top view represents surface (V) ?
17. What is the overall height of the shuttle?
18. Determine distances (20) and (23) .
19. What is the extreme length of the shuttle?
20. What is the distance between lines (5) and (21)?
21. What line in the front view represents surfaces (L) ?
22. Calculate distance (25) .

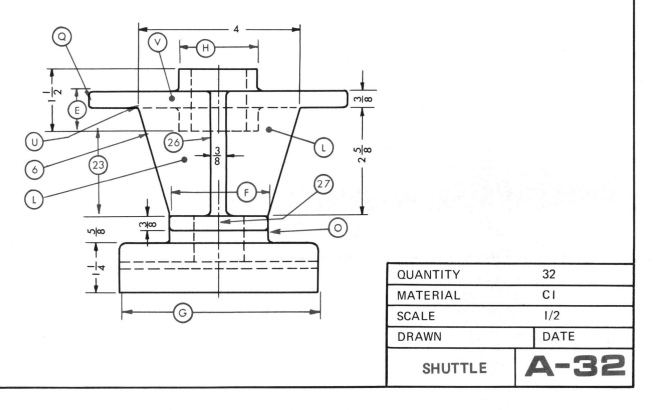

QUANTITY	32
MATERIAL	C I
SCALE	1/2
DRAWN	DATE
SHUTTLE	A-32

AUXILIARY VIEWS

Many machine parts have surfaces that are not perpendicular or at right angles to the plane of projection. These surfaces are referred to as sloping or inclining surfaces. In the regular orthographic views such surfaces appear to be foreshortened, and their true shape is not shown. When an inclined surface has important characteristics that should be shown very clearly and without distortion, an auxiliary view is used so that the drawing explains completely and clearly the shape of the object.

For example, figure 20-1 clearly shows why an auxiliary view is required. The circular features on the sloped surface on the front view cannot be seen in their true shape on either the top or side views. The auxiliary view is the only view which shows the true shape of these features. Note that only the sloped surface details are shown. Background detail is often omitted on auxiliary views and regular views to simplify the drawing and to avoid confusion. A break line is used to signify the break in an incomplete view. The break line is not required if only the exact surface is drawn for either an auxiliary view or a partial regular view. Dimensions for the detail on the inclined face are placed on the auxiliary view, where such a detail is seen in its true shape.

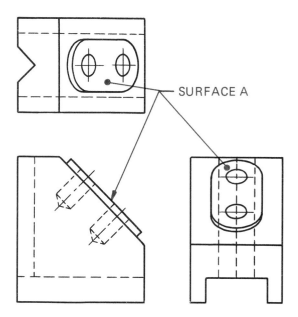

(A) REGULAR VIEWS DO NOT SHOW TRUE FEATURES OF SURFACE A.

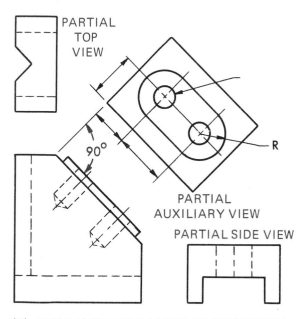

(B) AUXILIARY VIEW ADDED TO SHOW TRUE FEATURES OF SURFACE A.

Fig. 20-1 The Need for Auxiliary Views.

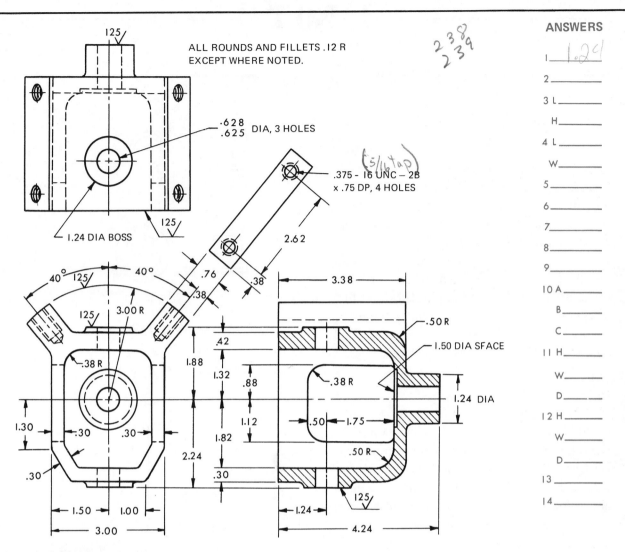

ALL ROUNDS AND FILLETS .12 R
EXCEPT WHERE NOTED.

.628
.625 DIA, 3 HOLES

1.24 DIA BOSS

.375 - 16 UNC – 2B
x .75 DP, 4 HOLES
(5/16 tap)

2.62

40°   125   40°

3.00 R

125

.76
.38
.38

3.38

.50 R

1.50 DIA SFACE

.38 R

1.24 DIA

.42
1.88
1.32   .88

.50   1.75

1.12
1.82

.38 R

1.30   .30   .30

2.24   .30

.30

1.50   1.00

3.00

.50 R

125

1.24

4.24

ANSWERS

1 ___ 1.24
2 _____
3 L _____
H _____
4 L _____
W _____
5 _____
6 _____
7 _____
8 _____
9 _____
10 A _____
B _____
C _____
11 H _____
W _____
D _____
12 H _____
W _____
D _____
13 _____
14 _____

QUESTIONS

1. What is the diameter of the bosses?

2. What tolerance is given on the holes in the bosses?

3. What is the length and height of the cutouts in the sides of the box?

4. What is the length and width of the legs of the box?

5. How many degrees are there between the legs of the box?

6. What is the maximum surface roughness permitted on the machined surface?

7. How many surfaces are machined?

8. What is the inside diameter of the pipe that this box fits into?

9. What size of cap screws would be used to fasten the box to the pipe?

10. What is the thickness of (A) the side walls of the box; (B) the top of the box; (C) the bottom of the box?

11. Give overall inside dimensions of the box.

12. What are the overall outside dimensions of the box? (Do not include legs or bosses.)

13. Of what material is the box cast?

14. How many screws are required to fasten the box to the pipe?

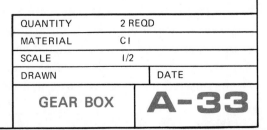

QUANTITY	2 REQD	
MATERIAL	C I	
SCALE	1/2	
DRAWN		DATE

GEAR BOX   **A-33**

103

UNIT 21

BEARINGS [1]

All bearings, both plain and rolling elements, have their own particular advantages and disadvantages. No universally excellent bearing exists. Bearings permit relative and controlled motion between two loaded members. From a practical point of view, a bearing must support an applied load over a given range of speeds for a minimum specified life. The motion may involve rolling elements or sliding members. Either system may use fluid films, boundary films, or dry surfaces.

Plain Bearings [2]

Plain bearings are used in a wide variety of applications and have a wide variety of shapes and sizes. Because of their simplicity, they are very versatile. They may be categorized into several groups, and the most common are journal (sleeve) bearings and thrust bearings. These are available in a variety of standard sizes and shapes.

Journal or Sleeve Bearings. Cast bronze and porous bronze cylindrical bearings are the simplest and most widely used sleeve bearings. Cast bronze bearings are lubricated with grease or oil. Porous bearings are impregnated with oil and are often provided with a reservoir in the housing to replenish oil as it is depleted from the pores. These bearings are usually used in lightly loaded, low-speed applications where oil or grease lubrication is not desirable or possible.

Thrust Bearings. Plain thrust bearings or thrust washers can be obtained from manufacturers in a variety of materials including sintered metal, plastic, woven TFE fabric on steel backing, sintered teflon-bronze-lead on metal back-

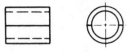

SLEEVE

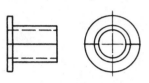

FLANGED
(A) JOURNAL TYPE

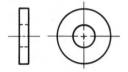

(B) THRUST TYPE

Fig. 21-1   Plain Bearings.

ing, aluminum alloy on steel, aluminum alloy, and carbon-graphite.

Boring of Split Bearings

In the design and construction of bearings it is sometimes necessary to split the bearing to facilitate assembling to permit adjusting, and also to make possible the replacing of worn parts. The split type of bearing allows the shaft or spindle to be set in one half of the bearing while the other half, or cover, is later secured in position. See figure 21-2.

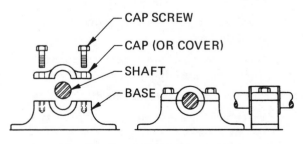

CAP SCREW
CAP (OR COVER)
SHAFT
BASE

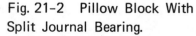

Fig. 21-2   Pillow Block With
Split Journal Bearing.

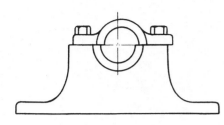

Fig. 21-3   Bearing Halves
Incorrectly Matched.

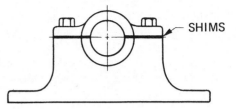

Fig. 21-4   "Shimmed" Bearing.

If the bearings shown in figure 21-2 are to be made from two parts, it is necessary that they be fastened together before the hole is bored or reamed. If this were not done, the machining operation would be more difficult and the two parts might not make a perfectly round bearing when assembled. See figure 21-3.

One method used to give longer life to the bearing is to insert very thin strips of metal between the base and cover halves before boring. These thin strips which vary in thickness are called *shims*.

When a bearing is shimmed, the same number of pieces of corresponding thickness are used on both sides of the bearing.

As the hole wears, one or more pairs of these shims may be removed to compensate for wear.

REFERENCES AND SOURCE MATERIAL

1. A.O. De Hart, "Basic Bearing Types" *Machine Design,* 40, No. 14 (1968).

2. W.A. Glaeser, "Plain Bearings" *Machine Design,* 40, No. 14 (1968).

SKETCHING ASSIGNMENT: MAKE A FREEHAND SKETCH
IN THE GRAPH AREA SHOWN BELOW, SHOWING THE
AUXILIARY VIEW OF THE SPLIT BEARING SURFACES (K)
AND (G).   SEE QUESTIONS 16 AND 17.

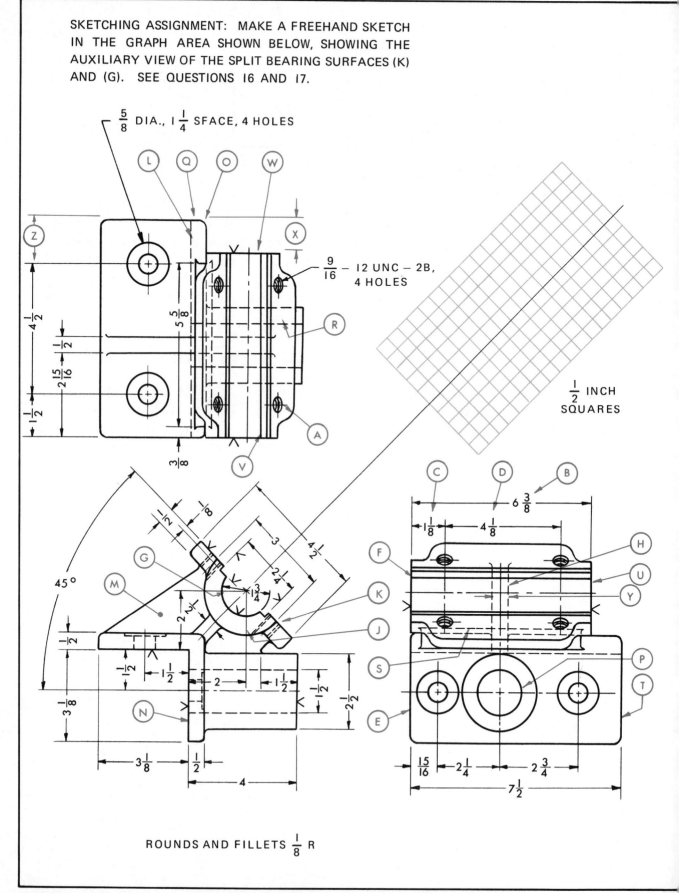

$\frac{5}{8}$ DIA., $1\frac{1}{4}$ SPACE, 4 HOLES

$\frac{9}{16}$ − 12 UNC − 2B, 4 HOLES

$\frac{1}{2}$ INCH SQUARES

ROUNDS AND FILLETS $\frac{1}{8}$ R

ADAPTED FROM "MECHANICAL DRAWING," HENRY FORD TRADE SCHOOL, DEARBORN, MICHIGAN

QUESTIONS

I. How many definite finished surfaces are on the casting?

2. What is the number and size of the threaded holes?

3. What is the number and size of holes bored?

4. What is the number and size of the mounting holes?

5. Note that surface (E) is not finished, but surface (F) is to be finished. Allowing 1/8" for finishing, how long will the rough casting be in the side view?

6. Would hole (G) be bored before or after bearing cap is assembled?

7. What line in the side view shows surface (M) ?

8. In which view, and by what line, is the surface represented by line (N) shown?

9. What line or surface in the side view shows the projection of point (J) ?

10. What point or surface in the side view does line (R) represent?

II. Locate surface (T) in the top view.

12. Locate surface (U) in the top view.

13. Locate (V) in the side view.

14. What are dimensions (X)(Y)(Z) ?

15. What is the size of the round (O) ?

16. Determine dimension (A) and place it correctly on the sketch of the auxiliary view.

17. Place dimensions (B)(C) and (D) correctly on the sketch of the auxiliary view.

ANSWERS

I _____

2 _____

3 _____

4 _____

5 _____

6 _____

7 _____

8 _____

9 _____

10 _____

II _____

12 _____

13 _____

14 (X)_____

(Y)_____

(Z)_____

15 _____

16 _____

QUANTITY	36	
MATERIAL	C I	
SCALE	1/3	
DRAWN		DATE
CORNER BRACKET	**A-34**	

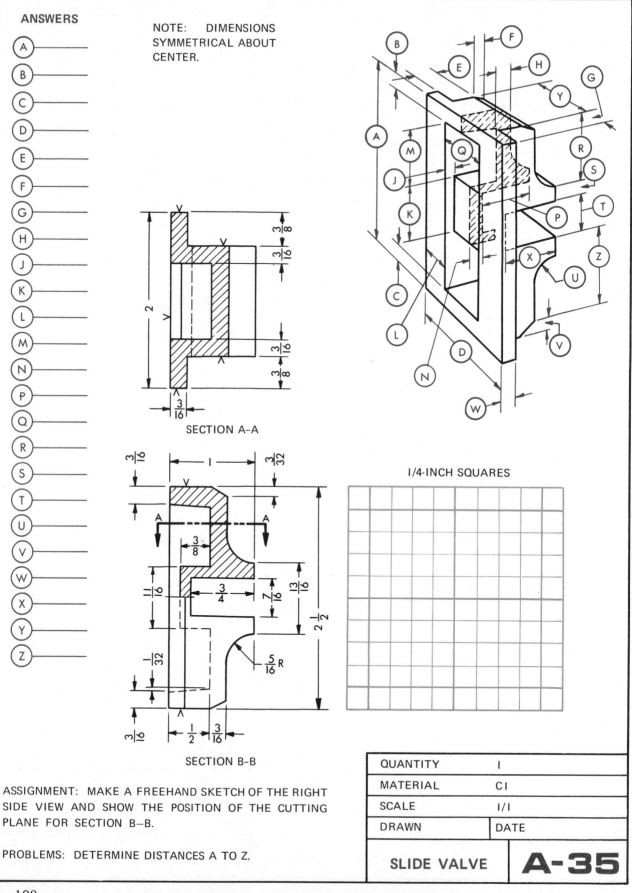

ANSWERS

A
B
C
D
E
F
G
H
J
K
L
M
N
P
Q
R
S
T
U
V
W
X
Y
Z

NOTE: DIMENSIONS SYMMETRICAL ABOUT CENTER.

SECTION A-A

SECTION B-B

1/4-INCH SQUARES

ASSIGNMENT: MAKE A FREEHAND SKETCH OF THE RIGHT SIDE VIEW AND SHOW THE POSITION OF THE CUTTING PLANE FOR SECTION B–B.

PROBLEMS: DETERMINE DISTANCES A TO Z.

QUANTITY	I
MATERIAL	CI
SCALE	I/I
DRAWN	DATE

SLIDE VALVE | **A-35**

108

UNIT 22

SIMPLIFIED DRAFTING

The challenge of modern industry is to produce more and better goods at competitive prices. Drafting, like all other branches of industry, must share in the responsibility for making this increased productivity possible. The old concept of drafting, that of producing an elaborate and beautiful drawing, complete with all the lines, projected views, and sections, must give way to a simplified method of drafting. This new simplified method of drafting must embrace many modern economical drafting practices but surrender nothing in either *clarity* of presentation or *accuracy* of dimensions. Drafting stripped of its frills is the new standard.

When a large number of holes of similar size are to be made in a part, there may be a chance that the man producing the part may misinterpret a conventional drawing as shown in figure 22-3A. To simplify the drawing and reduce the chance of errors, simplified drafting practices such as those shown in figure 22-3B are recommended.

Arrowless dimensioning removes many of the lines which clutter the drawing, and hole symbols simplify the location of similarly sized holes.

The use of the abbreviation ₵ indicates that all dimensions are symmetrical about the line indicated.

Hole Symbols

The symbols shown in figure 22-1 are recommended to designate various hole sizes on the drawing. The method of producing

holes should not be specified on the drawing because the holes should be made to suit factory planning.

On parts such as tubes the drawing should state "near side only," "far side only," or "thru holes" to eliminate any doubt on the part of the factory operator.

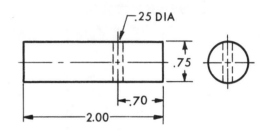

(A) CONVENTIONAL DRAWING

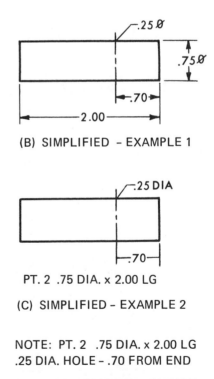

(B) SIMPLIFIED - EXAMPLE 1

PT. 2 .75 DIA. x 2.00 LG

(C) SIMPLIFIED - EXAMPLE 2

NOTE: PT. 2 .75 DIA. x 2.00 LG
.25 DIA. HOLE - .70 FROM END

(D) PART DESCRIBED BY NOTE
SIMPLIFIED - EXAMPLE 3

Fig. 22-2 Simplified Drafting Practices for Detailed Parts.

Fig. 22-1 Recommended Hole Symbols.

109

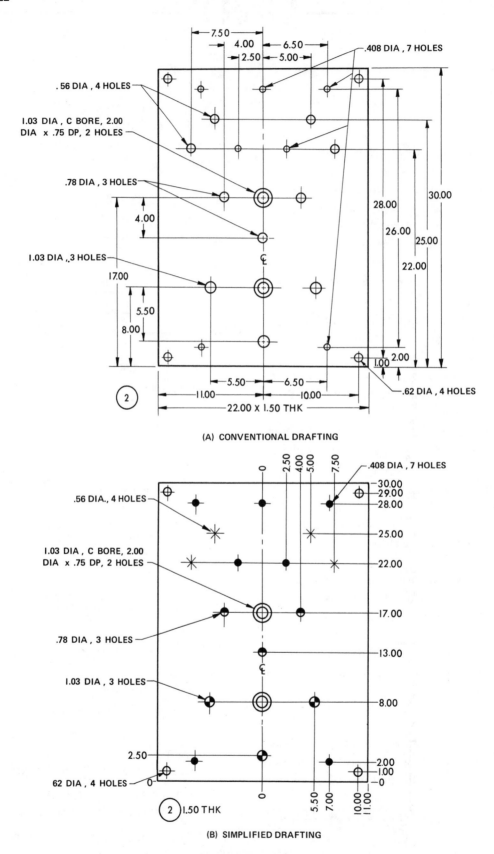

(A) CONVENTIONAL DRAFTING

(B) SIMPLIFIED DRAFTING

Fig. 22-3

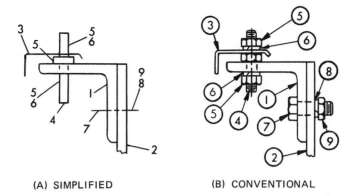

(A) SIMPLIFIED   (B) CONVENTIONAL

Fig. 22-4   Simplified Drafting Practices for Assembly Drawings.

Other Timesaving Practices

- Show only partial views of symmetrical objects.

- Avoid the use of elaborate pictorial or repetitive detail.

- Omit detail of nuts, bolt heads, and other hardware.

- Avoid the use of unnecessary hidden lines which do not add clarification.

- Use description wherever practical to eliminate drawings.

- Use symbols instead of words.

- Within limits, a small drawing is made more easily and quickly than a large drawing.

- Eliminate repetitive data by use of general notes.

- Omit part number circles and arrows on leader lines when it will not cause confusion with other data on the drawing.

- Omit centerlines except when necessary for processing.

- Eliminate views where the shape can be given by description; for example, hex, sq., dia, on $\mathcal{C}_L$, thk, etc.

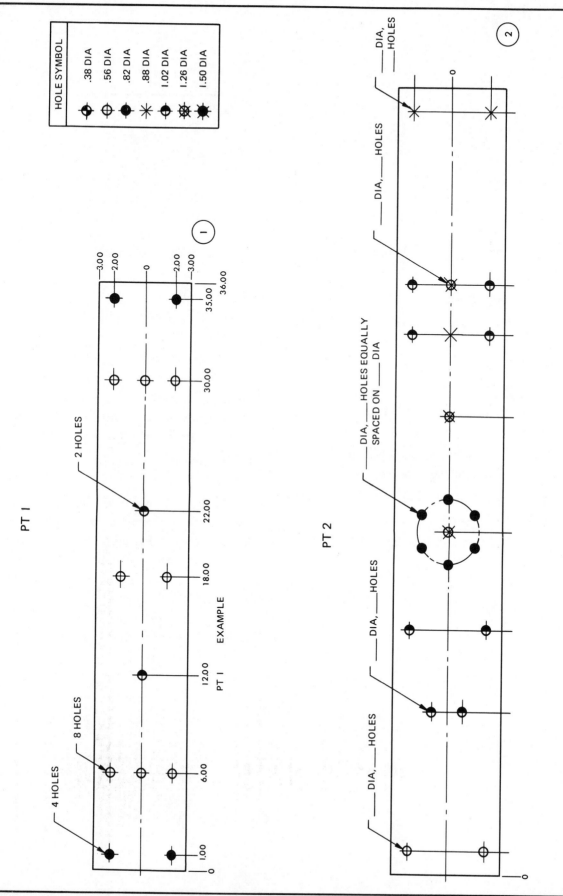

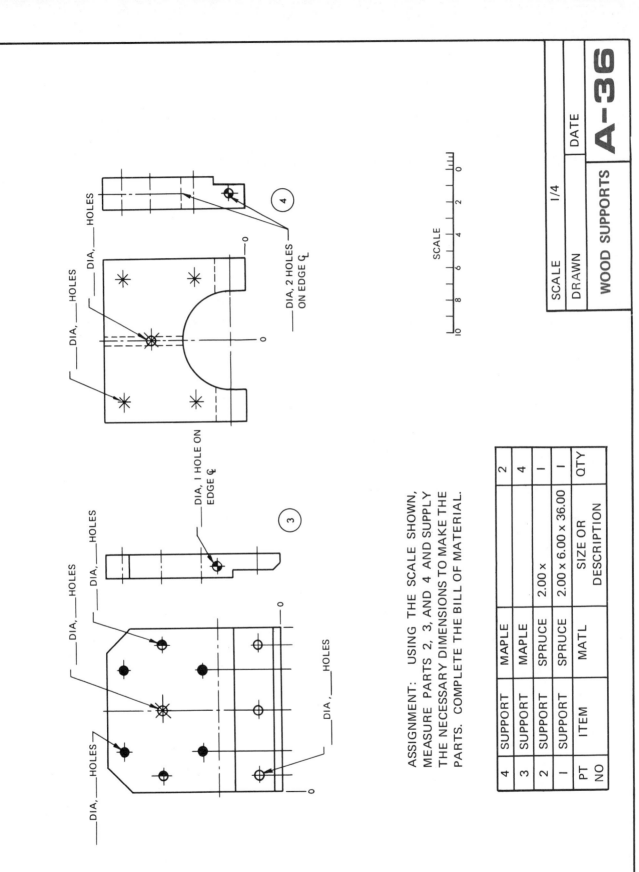

ASSIGNMENT: USING THE SCALE SHOWN, MEASURE PARTS 2, 3, AND 4 AND SUPPLY THE NECESSARY DIMENSIONS TO MAKE THE PARTS. COMPLETE THE BILL OF MATERIAL.

SCALE

4	SUPPORT	MAPLE		2
3	SUPPORT	MAPLE		4
2	SUPPORT	SPRUCE	2.00 ×	1
1	SUPPORT	SPRUCE	2.00 × 6.00 × 36.00	1
PT NO	ITEM	MATL	SIZE OR DESCRIPTION	QTY

SCALE	1/4	DATE
DRAWN		

WOOD SUPPORTS

A-36

113

ASSIGNMENT: IN THE GRAPH AREA PRO-
VIDED, SHOW THE SIMPLIFIED METHOD OF
DETAILING THE PARTS.

GRAPH SQUARE SIZE — .20 INCHES
(5 SQUARES TO THE INCH)

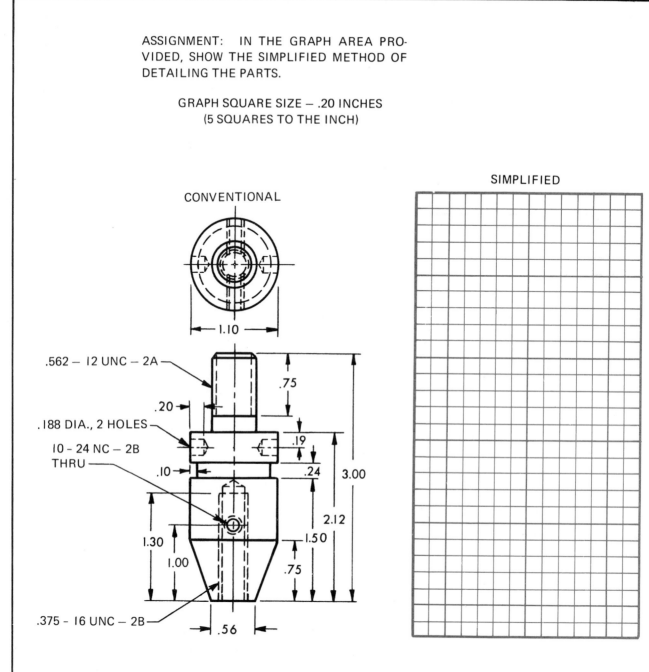

CONVENTIONAL

SIMPLIFIED

.562 — 12 UNC — 2A

.188 DIA., 2 HOLES

10 – 24 NC – 2B
THRU

.375 – 16 UNC – 2B

PT 1   COUPLING
MATL — SAE 1020 — 2 REQD

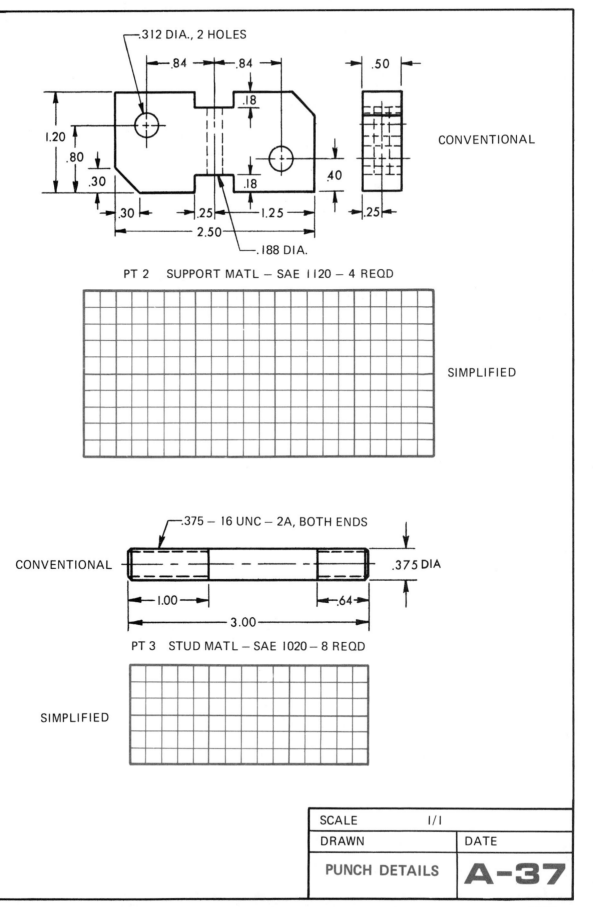

.312 DIA., 2 HOLES

.84 .84

.18

.50

CONVENTIONAL

1.20

.80

.18

.40

.30

.30 .25 1.25 .25

2.50

.188 DIA.

PT 2   SUPPORT MATL — SAE 1120 — 4 REQD

SIMPLIFIED

.375 — 16 UNC — 2A, BOTH ENDS

CONVENTIONAL

.375 DIA

1.00 .64

3.00

PT 3   STUD MATL — SAE 1020 — 8 REQD

SIMPLIFIED

SCALE	1/1	
DRAWN		DATE
PUNCH DETAILS		A-37

115

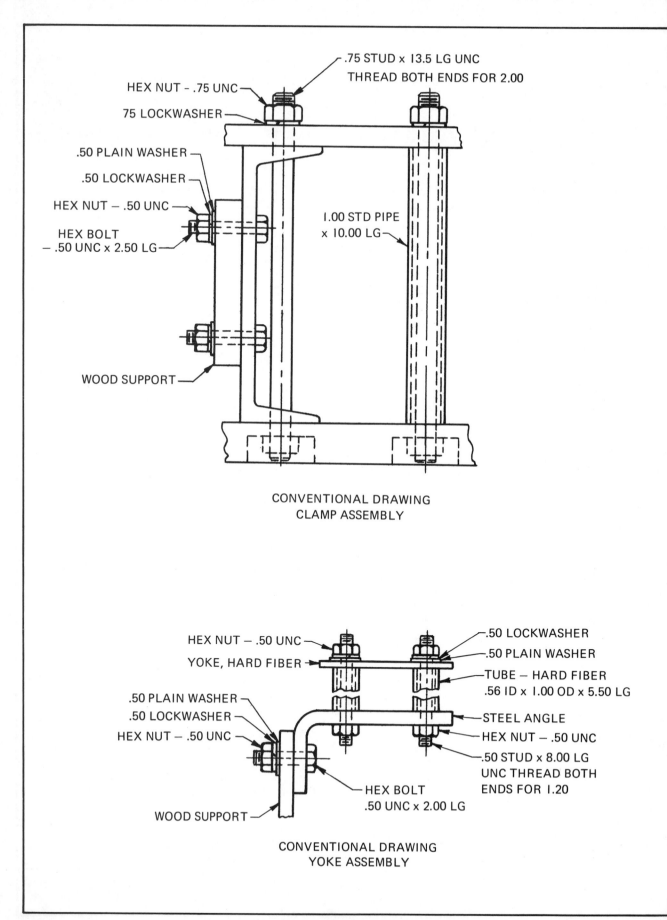

.75 STUD x 13.5 LG UNC
THREAD BOTH ENDS FOR 2.00

HEX NUT – .75 UNC

75 LOCKWASHER

.50 PLAIN WASHER

.50 LOCKWASHER

HEX NUT – .50 UNC

HEX BOLT
– .50 UNC x 2.50 LG

1.00 STD PIPE
x 10.00 LG

WOOD SUPPORT

CONVENTIONAL DRAWING
CLAMP ASSEMBLY

HEX NUT – .50 UNC

YOKE, HARD FIBER

.50 LOCKWASHER

.50 PLAIN WASHER

TUBE – HARD FIBER
.56 ID x 1.00 OD x 5.50 LG

.50 PLAIN WASHER

.50 LOCKWASHER

HEX NUT – .50 UNC

STEEL ANGLE

HEX NUT – .50 UNC

.50 STUD x 8.00 LG
UNC THREAD BOTH
ENDS FOR 1.20

HEX BOLT
.50 UNC x 2.00 LG

WOOD SUPPORT

CONVENTIONAL DRAWING
YOKE ASSEMBLY

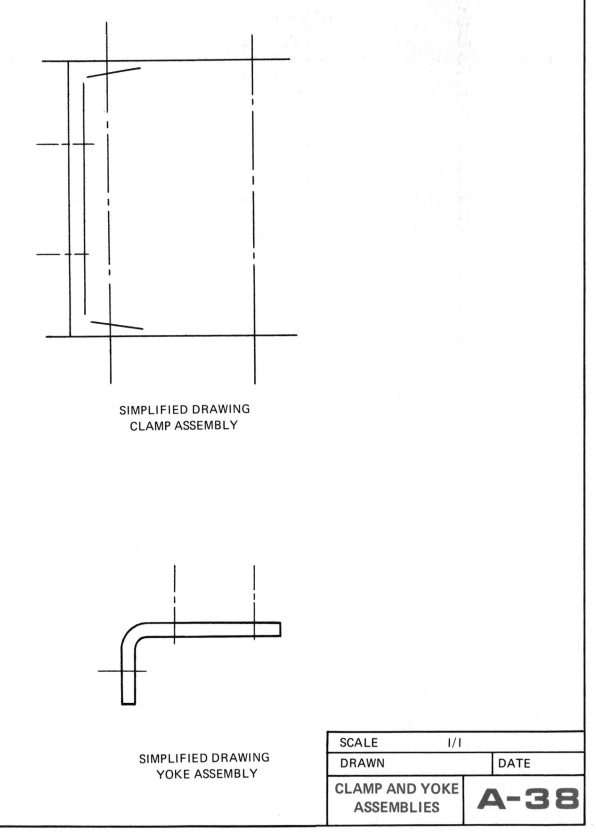

SIMPLIFIED DRAWING
CLAMP ASSEMBLY

SIMPLIFIED DRAWING
YOKE ASSEMBLY

SCALE	1/1	
DRAWN		DATE
CLAMP AND YOKE ASSEMBLIES	A-38	

ARRANGEMENT OF VIEWS

The shape of an object and its complexity influence the possible choices and arrangement of views of that particular part. Since one of the main purposes of making drawings is to furnish the mechanic with enough information to be able to make the object, only those views should be drawn which will aid in the interpretation of the drawing.

Ordinarily, the part being drawn has for its front view that view which, in the opinion of the draftsman, gives the best idea of the purpose and general contour of the object. This choice of the front view may have no relationship to the actual front of the piece when it is mounted on the machine. In other words, the *front view* may not be the *actual front* of the object itself.

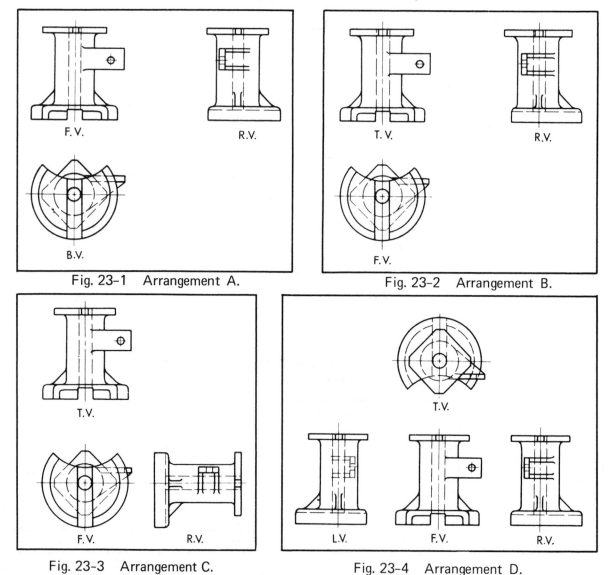

Fig. 23-1   Arrangement A.

Fig. 23-2   Arrangement B.

Fig. 23-3   Arrangement C.

Fig. 23-4   Arrangement D.

The Index Pedestal, which follows, is an example of an object of such a shape that it is desirable to draw the views as they are shown on drawing A-39. These views could be called front, right, and bottom views, as illustrated in figure 23-1. These views could also be designated as: front, top, and right side, as shown in figure 23-2. However, the designation is not of major importance. What is important is that these views, in the opinion of the draftsman or designer, give the necessary information in the way easiest to understand. The views may also be arranged as shown in figure 23-3 or 23-4. While these arrangements would be acceptable, they are not quite as easy to read.

DRILL SIZES

Twist drills are the most common tools used in drilling and are made in many sizes. They are grouped according to number sizes, 1 to 80, corresponding to the Stubbs steel wire gage; by letter sizes A to Z; by fractional sizes 1/64 up or their decimal equivalents; and by millimeter sizes. Number and letter size drills are listed in Table 6 of the Appendix.

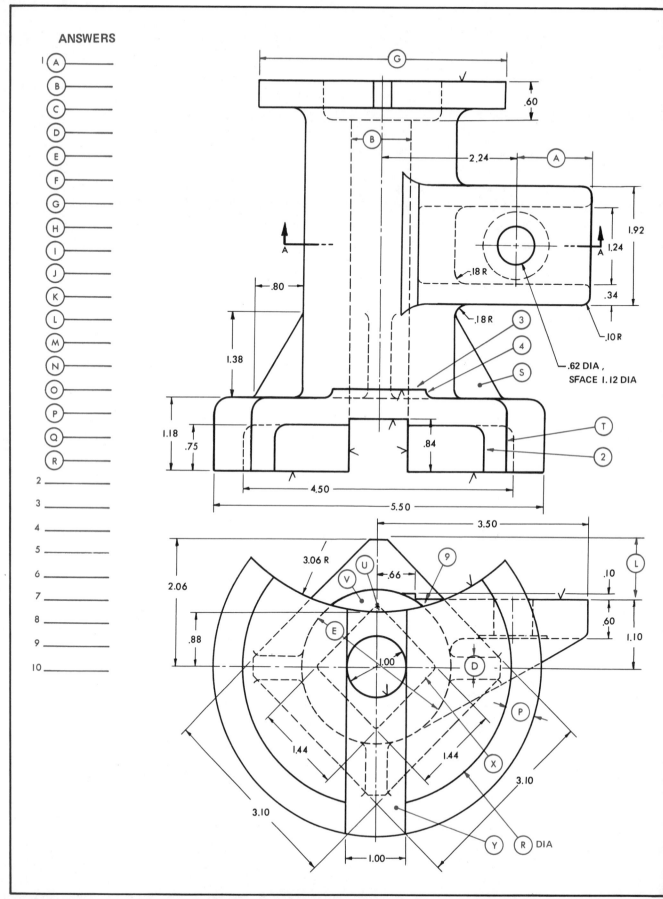

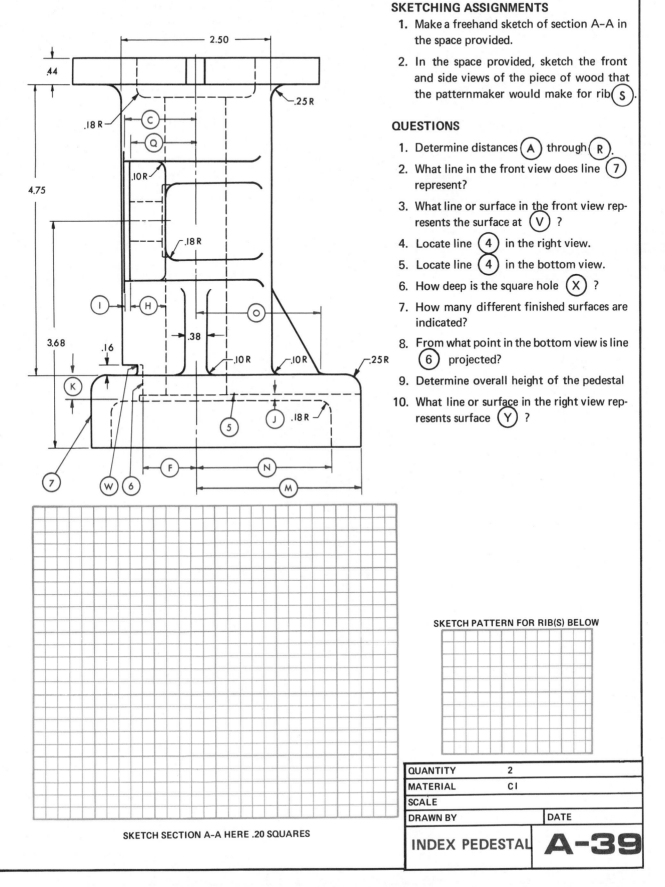

SKETCHING ASSIGNMENTS

1. Make a freehand sketch of section A–A in the space provided.

2. In the space provided, sketch the front and side views of the piece of wood that the patternmaker would make for rib (S).

QUESTIONS

1. Determine distances (A) through (R).

2. What line in the front view does line (7) represent?

3. What line or surface in the front view represents the surface at (V) ?

4. Locate line (4) in the right view.

5. Locate line (4) in the bottom view.

6. How deep is the square hole (X) ?

7. How many different finished surfaces are indicated?

8. From what point in the bottom view is line (6) projected?

9. Determine overall height of the pedestal

10. What line or surface in the right view represents surface (Y) ?

SKETCH SECTION A-A HERE .20 SQUARES

SKETCH PATTERN FOR RIB(S) BELOW

QUANTITY	2	
MATERIAL	C I	
SCALE		
DRAWN BY		DATE

INDEX PEDESTAL | **A-39**

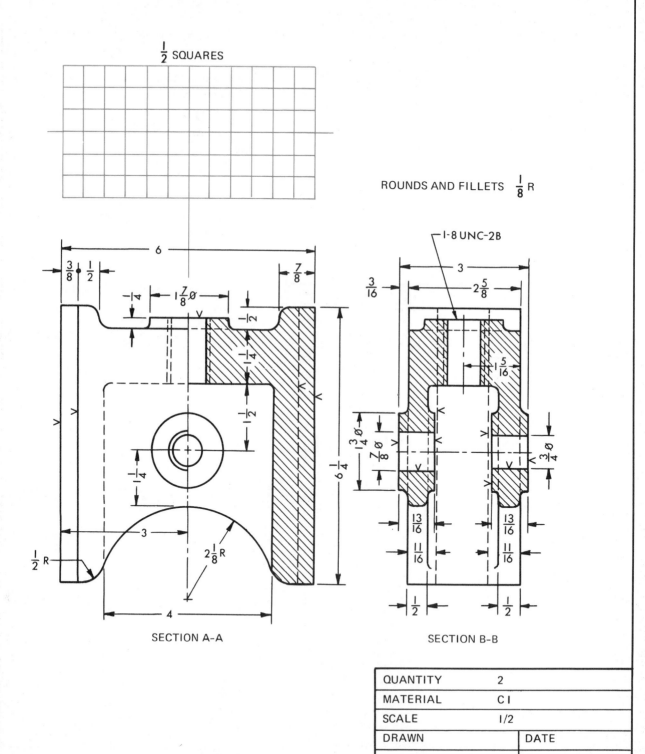

ASSIGNMENT: MAKE A FREEHAND SKETCH OF THE TOP VIEW OF THE CROSS HEAD IN THE SPACE PROVIDED. SEVERAL DESIGNS ARE POSSIBLE. SHOW THE POSITIONS OF CUTTING LINES A-A AND B-B ON THE TOP VIEW.

$\frac{1}{2}$ SQUARES

ROUNDS AND FILLETS $\frac{1}{8}$ R

1-8 UNC-2B

SECTION A-A

SECTION B-B

QUANTITY	2	
MATERIAL	C I	
SCALE	1/2	
DRAWN		DATE
CROSS HEAD		**A-40**

UNIT 24

ALIGNMENT OF PARTS AND HOLES

Two of the important factors to be considered in the drawing of a machine part are: the number of views to be drawn and the time required to draw them. If it is possible to use some timesaving device and less space for making the drawing, consistent with clearness and easy reading, then such practice is desirable.

To simplify the representation of common features, a number of conventional drawing practices are used. Many conventions are deviations from true projection for the purpose of clarity; others are used for the purpose of saving drafting time. These conventions must be executed carefully, for clarity is even more important than speed.

Foreshortened Projection

When the true projection of a piece would result in confusing foreshortening, parts such as ribs or arms should be rotated until parallel to the line of the section or projection.

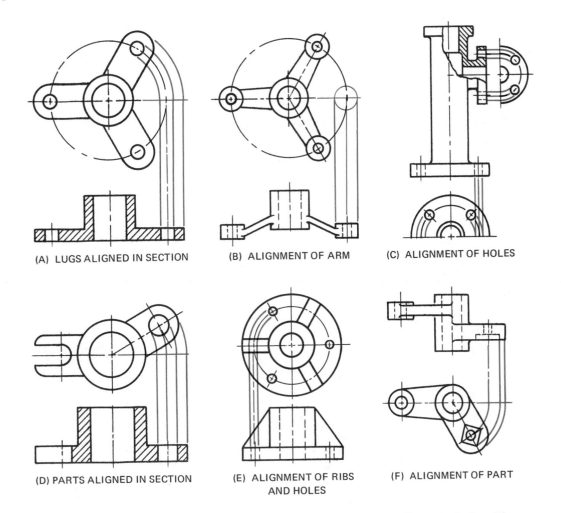

(A) LUGS ALIGNED IN SECTION   (B) ALIGNMENT OF ARM   (C) ALIGNMENT OF HOLES

(D) PARTS ALIGNED IN SECTION   (E) ALIGNMENT OF RIBS AND HOLES   (F) ALIGNMENT OF PART

Fig. 24-1 Alignment of Holes and Parts to Show Their True Relationship.

Holes Revolved to Show True Center Distance

Drilled flanges in elevation or section should show the holes at their true distance from center, rather than the true projection.

NAMING OF VIEWS FOR SPARK ADJUSTER

The drawing of the spark adjuster, drawing A-41, illustrates several violations of true projection. There could be some question about the names of the views of the spark adjuster. The importance lies not in the names but rather in the relationship of the views to each other. This means that the right view must be on the right side of the front, the left view must be on the left side of the front view, etc. Some confusion may result in naming the views of the spark adjuster. Any combination in figure 24-2 may be used.

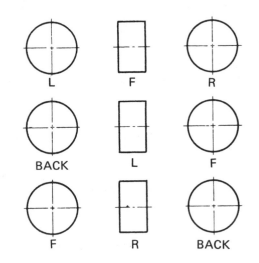

Fig. 24-2 Naming of Views for Spark Adjuster, Drawing A-41.

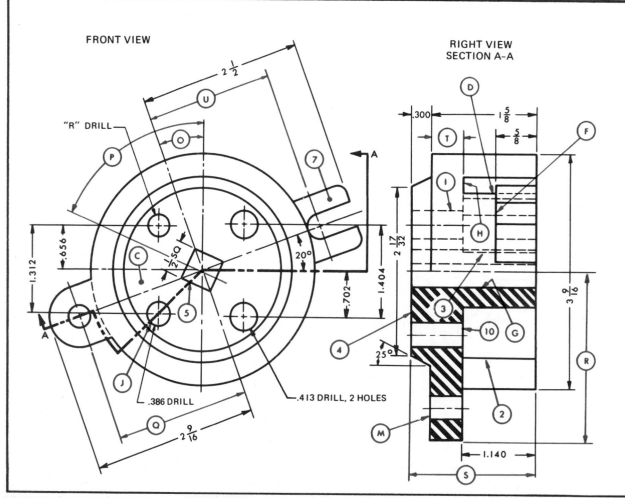

QUESTIONS

1. What is radius (A) ?
2. What surface is line (2) in the back view?
3. Locate surface (B) in section A-A.
4. Locate surface (C) in section A-A.
5. Locate line (D) in the back view.
6. Locate a surface or line in the front view that represents line (F) .
7. What point or line in the front view does line (G) represent?
8. What letter drill would be used to drill hole (J) ?
9. What drill would be used to drill hole (K) ?
10. Locate point or line in the back view from which line (I) is projected.
11. Locate surface (H) in the back view.
12. Locate (Z) in section A-A;
13. How thick is lug (7) ?
14. What are the diameters of holes (L)(M)(N) ?
15. Determine angles (O)(P) .
16. Determine distances (Q) through (U) .

ANSWERS

1 _____
2 _____
3 _____
4 _____
5 _____
6 _____
7 _____
8 _____
9 _____
10 _____
11 _____
12 _____
13 _____
14 (L) _____
(M) _____
(N) _____
15 (O) _____
(P) _____
16 (Q) _____
(R) _____
(S) _____
(T) _____
(U) _____

BACK VIEW

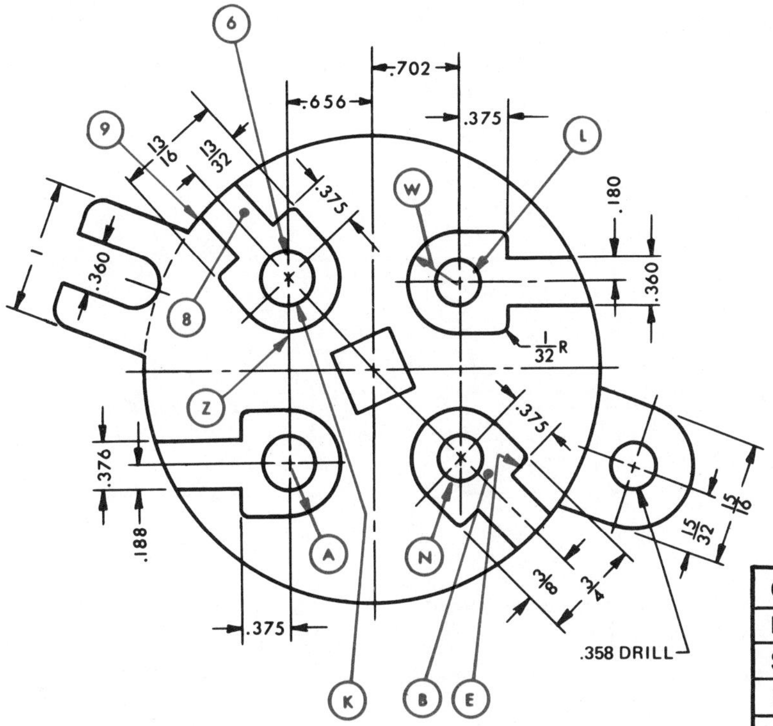

QUANTITY	6
MATERIAL	BAKELITE
SCALE	1/1
DRAWN	DATE

SPARK ADJUSTER

A-41

125

UNIT 25

BROKEN OUT AND PARTIAL SECTIONS

Broken-out and partial sections are used when it is desirable to show certain internal and external features of an object without drawing another view. A cutting plane line or a broken line is used to indicate where the section is taken. In the front view of the next assignment, the raise block, drawing A-42, two partial sections are used. Although this method of showing a partial section is somewhat unusual, it is accepted practice.

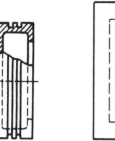

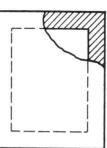

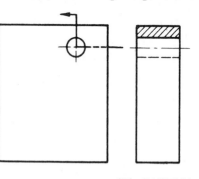

(A) BROKEN-OUT SECTIONS

(B) PARTIAL SECTION

Fig. 25-1   Broken-Out and Partial Sections.

QUESTIONS

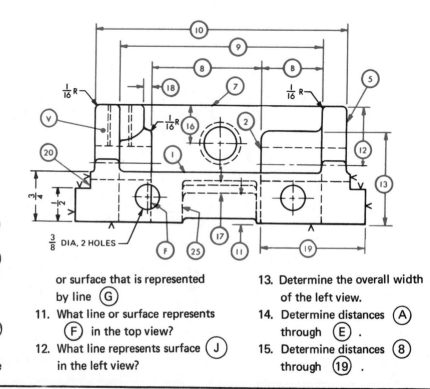

1. What is the diameter of the largest hole (not threaded)?
2. What is the size of the smallest threaded hole?
3. Give the number and the size of the smallest drilled holes.
4. What surface does ⑦ represent in the top view?
5. What surface does ① represent in the top view?
6. What line or surface does ⑥ represent in the left view?
7. What line or surface does Ⓥ represent in the front view?
8. How many finished surfaces are indicated?
9. By what line or surface is Ⓗ represented in the left view?
10. Locate in the left view the line or surface that is represented by line Ⓖ
11. What line or surface represents Ⓕ in the top view?
12. What line represents surface Ⓙ in the left view?
13. Determine the overall width of the left view.
14. Determine distances Ⓐ through Ⓔ .
15. Determine distances ⑧ through ⑲ .

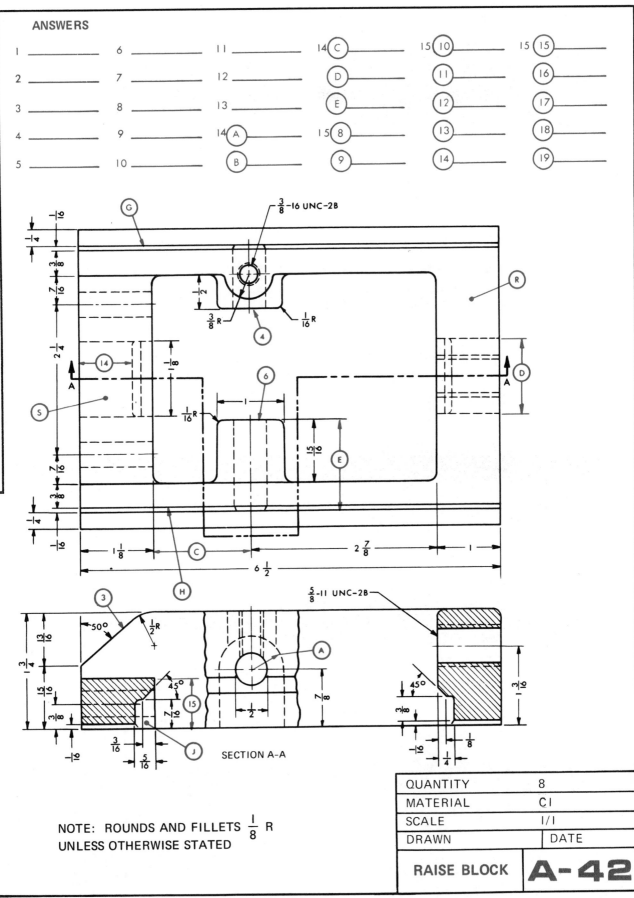

ANSWERS

1 _____	6 _____	11 _____	14 Ⓒ _____	15 ⑩ _____	15 ⑮ _____
2 _____	7 _____	12 _____	Ⓓ _____	⑪ _____	⑯ _____
3 _____	8 _____	13 _____	Ⓔ _____	⑫ _____	⑰ _____
4 _____	9 _____	14 Ⓐ _____	15 ⑧ _____	⑬ _____	⑱ _____
5 _____	10 _____	Ⓑ _____	⑨ _____	⑭ _____	⑲ _____

$\frac{3}{8}$–16 UNC–2B

SECTION A–A

NOTE: ROUNDS AND FILLETS $\frac{1}{8}$ R
UNLESS OTHERWISE STATED

$\frac{5}{8}$–11 UNC–2B

QUANTITY	8
MATERIAL	CI
SCALE	1/1
DRAWN	DATE

RAISE BLOCK | A-42

127

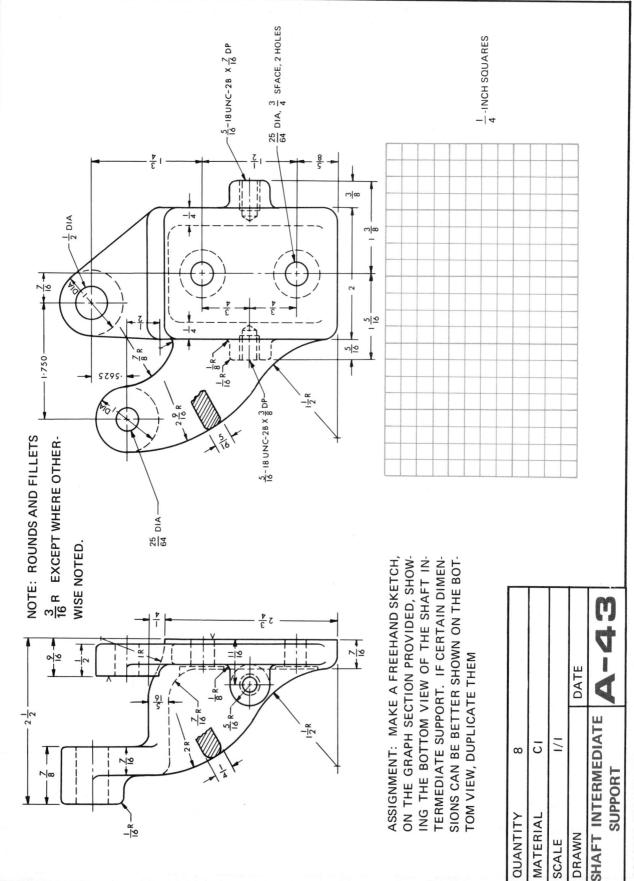

NOTE: ROUNDS AND FILLETS $\frac{3}{16}$ R EXCEPT WHERE OTHERWISE NOTED.

$\frac{5}{16}$-18UNC-2B X $\frac{7}{16}$ DP

$\frac{25}{64}$ DIA, $\frac{3}{4}$ SFACE, 2 HOLES

$\frac{1}{4}$-INCH SQUARES

$\frac{1}{2}$ DIA

$\frac{25}{64}$ DIA

$\frac{5}{16}$-18 UNC-2B X $\frac{3}{8}$ DP

ASSIGNMENT: MAKE A FREEHAND SKETCH, ON THE GRAPH SECTION PROVIDED, SHOWING THE BOTTOM VIEW OF THE SHAFT INTERMEDIATE SUPPORT. IF CERTAIN DIMENSIONS CAN BE BETTER SHOWN ON THE BOTTOM VIEW, DUPLICATE THEM

QUANTITY	8
MATERIAL	C I
SCALE	I/I
DRAWN	DATE

SHAFT INTERMEDIATE SUPPORT

A-43

ADAPTED FROM MACHINE DRAWING BY ELIOT F. TOZER AND HARRY A RISING, McGRAW-HILL BOOK COMPANY, INC., NEW YORK

CORED CASTINGS

Cored castings have certain advantages over solid castings. Where practical, castings are designed with cored holes or openings for the purposes of economy, appearance, and accessibility to interior surfaces.

The appearance of a casting is often improved by cored openings. In most instances, cored castings are more economical than solid castings due to the saving in metal. While cored castings are lighter, they are so designed that strength is not sacrificed. Unnecessary machining is also eliminated because of the openings that are cast in the part.

Hand holes may also be formed by coring in order to provide an opening through which the interior of the casting may be reached. These openings also permit machining an otherwise inaccessible surface of the part as shown in figure 26-1.

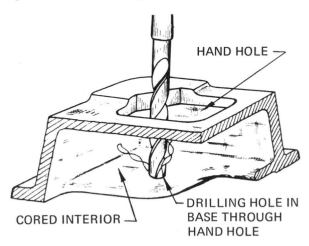

Fig. 26-1   Section of Cored Casting.

In the case of the auxiliary pump base, drawing A-44, the body of the casting is cored and provided with openings. Some of the reasons for coring this particular part are:

- *Economy:* The casting is lightened to save metal and weight.

- *Accessibility to interior:* The coring provides openings so that the four legs on the bottom of the base may be drilled and spotfaced on the inside. The hand holes also provide accessibility for reaching into the casting.

- *Appearance:* The object is designed to look more attractive.

SCRAPING

Scraping is an operation which is performed on machined surfaces in order to produce a surface which is flat, true, and smooth. Scraping is also used occasionally to correct small errors and provide for precision fits.

With the increased accuracy of modern machining methods, the need for scraping surfaces is becoming less and less. However, when scraping is necessary, it means that all *high spots* are to be removed and the surface made as smooth and accurate as possible.

The directions for this operation will be indicated on the drawing by an arrow pointing to the surface and the word *scrape.*

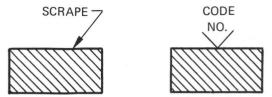

Fig. 26-2   Indicating Surfaces to be Scraped.

Another way of indicating a scraping operation is by using a code number, which is inserted above the symbol.

As a result of scraping, the finished surface has a frosted appearance. Machined surfaces are sometimes scraped to produce a decorative scroll or patchwork design.

129

1. Of what material is the base made?

2. How many finished surfaces are indicated?

3. What encircled letters on the drawing indicate the spaces that were cored when the casting was made?

4. Locate the surface in the top view that is represented by line (O) .

5. What is the height of the cored area (X) ?

6. What is the approximate length of the large cored area (X) ?

7. Give horizontal length of pads (2)(3)(4)(5) ?

8. What reason might be given for providing the openings (Q)(S)(Y)(Z) ?

9. Give the approximate distance for (W) .

10. What radius fillet would be used at (R) ?

11. Determine distances (C) through (N) .

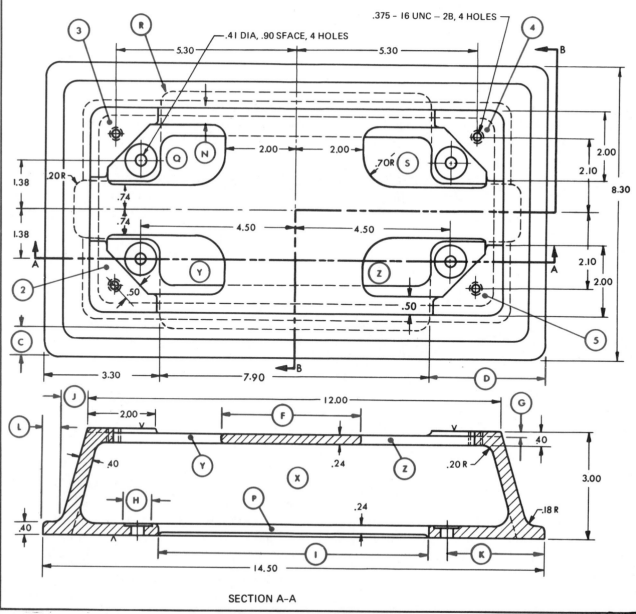

SECTION A-A

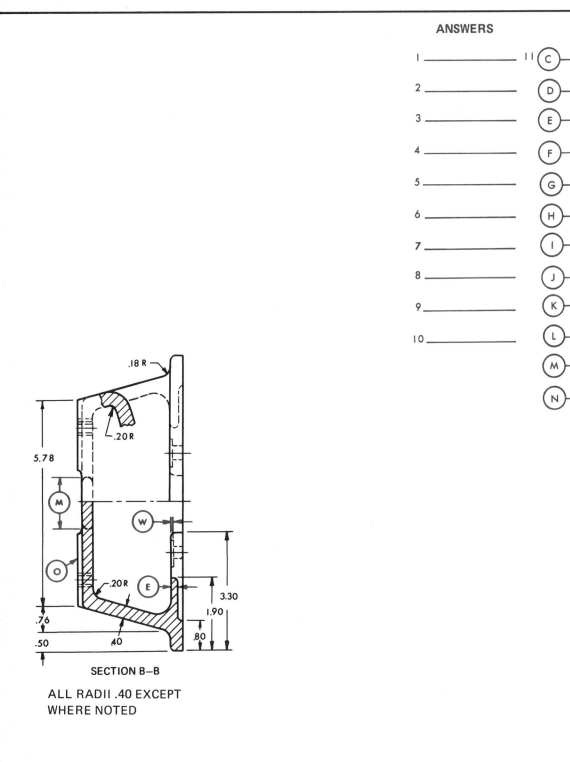

SECTION B–B

ALL RADII .40 EXCEPT
WHERE NOTED

NOTE: ALL DIMENSIONS SYMMETRICAL
ABOUT THE CENTERLINES.

QUANTITY	4	
MATERIAL	CI	
SCALE	1/2	
DRAWN		DATE
AUXILIARY PUMP BASE	**A-44**	

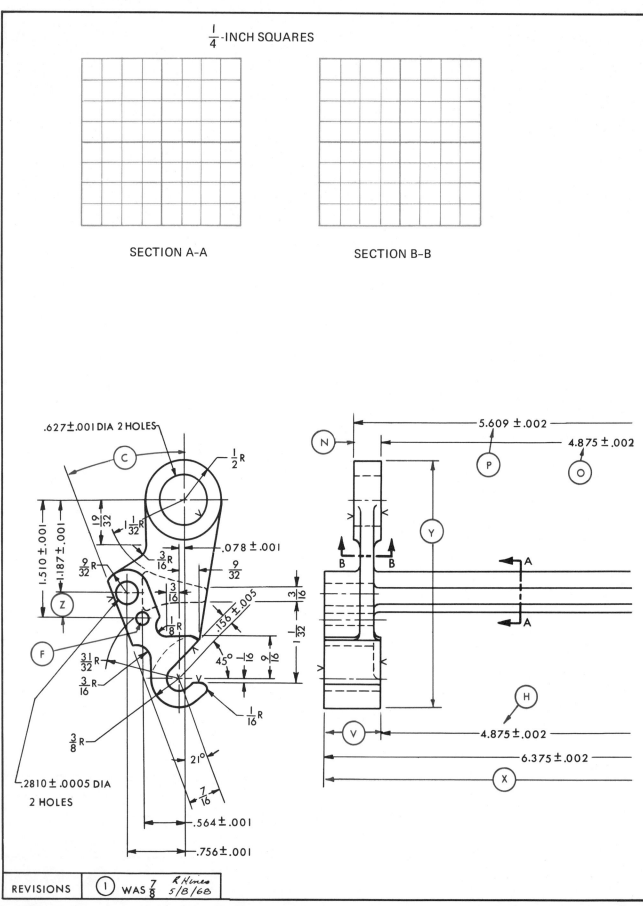

$\frac{1}{4}$-INCH SQUARES

SECTION A-A

SECTION B-B

.627±.001 DIA 2 HOLES

C

$\frac{1}{2}$ R

$\frac{19}{32}$

$1\frac{1}{32}$ R

.078 ±.001

$\frac{3}{16}$ R

$\frac{9}{32}$

$\frac{9}{32}$ R

$\frac{3}{16}$

1.510 ±.001

1.187 ±.001

Z

.156 ±.005

$\frac{3}{16}$

$\frac{1}{32}$

$\frac{1}{8}$ R

F

$\frac{31}{32}$ R

$\frac{3}{16}$ R

45° $\frac{1}{16}$

$\frac{9}{16}$

$1\frac{1}{32}$

$\frac{1}{16}$ R

$\frac{3}{8}$ R

21°

.2810 ±.0005 DIA
2 HOLES

$\frac{7}{16}$

.564±.001

.756±.001

N

5.609 ±.002

4.875 ±.002

P

O

B   B

Y

A

A

H

V

4.875 ±.002

6.375 ±.002

X

REVISIONS   (1) WAS $\frac{7}{8}$ R Hines   5/8/68

SKETCHING ASSIGNMENT

Make freehand sketches of sections A–A and B–B in the spaces provided. Assume all corners to have a 1/16-in. radius. Assume that these sections are part of the original drawing dimension where necessary.

QUESTIONS

1. State the number of definite surfaces that are to be finished.
2. Determine angles (C) and (G)
3. What are the sizes of holes (D)(E)(F) ?

4. What are the maximum acceptable dimensions for (V) , (I) if the 6.375 dimension is made to its maximum limit?
5. If dimension (L) is made exactly .564 and dimension (M) .757, what would dimension (K) be.
6. Determine distance (N) , if distances (Q)(O)(P) are made .001 over given sizes.
7. What circled number indicates a changed dimension?
8. What is radius (J) ?
9. Determine distances (S) to (Z).

ANSWERS

1 _____   3 (E) _____   5 _____   9 (S) _____   (W) _____

2 (C) _____   (F) _____   6 _____   (T) _____   (X) _____

(G) _____   4 (V) _____   7 _____   (J) _____   (Y) _____

3 (D) _____   (I) _____   8 _____   (V) _____   (Z) _____

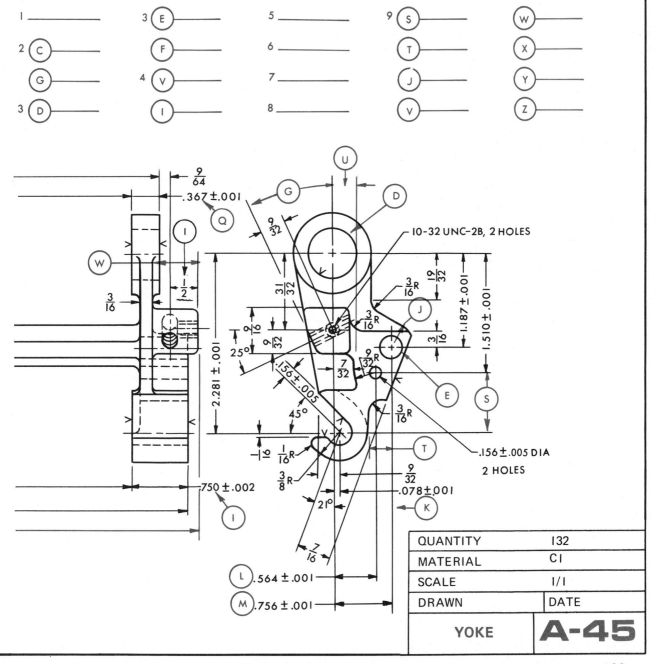

QUANTITY	132
MATERIAL	CI
SCALE	1/1
DRAWN	DATE

YOKE   **A-45**

133

UNIT 27

PIN FASTENERS [1]

Pin fasteners offer an inexpensive and effective approach to assembly where loading is primarily in shear. They can be conveniently divided into two groups.

- *Semipermanent:* Those pin fasteners that require application of pressure or the aid of tools for installation or removal. Rep-

resentative types include radial locking (grooved-surface, spring) and machine pins (taper, dowel, clevis, and cotter pins). Descriptive data can be found in figures 27-1 and 27-2.

- *Quick Release:* These are more elaborate self-contained pins which are used for rapid manual assembly or disassembly.

(A) SOLID WITH GROOVED SURFACES	
TYPE A	Full-length grooves. Used for general-purpose fastening.
TYPE B	Grooves extend half length of the pin. Used as a hinge or linkage "bolt" but also can be employed for other functions in through-drilled holes where a locking fit over only part of the pin length is required.
TYPE C	Full-length grooves with pilot section at one end to facilitate assembly. Expanded dimension of this pin is held to a maximum over the full-grooved length to provide uniform locking action. It is recommended for applications subject to severe vibration or shock loads where maximum locking effect is required.
TYPE D	Reverse tapered grooves extend half the pin length. It is the counterpart of the Type B pin for assembly in blind holes.
TYPE E	Half-length groove section centered along the pin surface. Used as a cotter pin or in similar functions where an artificial shoulder or a locking fit over the center portion of the pin is required.
TYPE F	Full-length grooves with pilot section at both ends for hopper feeding. Same as Type C.
(B) HOLLOW SPRING PINS	
SPIRAL-WRAPPED	
SLOTTED-TUBULAR	

Fig. 27-1 Radial Locking Pins. (Courtesy Machine Design, Vol. 39, No. 14, 1967)

	HARDENED AND GROUND DOWEL PIN: Standardized in nominal diameters ranging from 1/8″ to 7/8″. Used for: 1. Holding laminated sections together with surfaces either drawn up tight or separated in some fixed relationship. 2. Fastening machine parts where accuracy of alignment is a primary requisite. 3. Locking components on shafts, in the form of transverse pin key.
	TAPER PIN: Standard pins have a taper of 1/4″ per foot measured on the diameter. Basic dimension is the diameter of the large end. Used for light duty service in the attachment of wheels, levers and similar components to shafts. Torque capacity is determined on the basis of double shear, using the average diameter along the tapered section in the shaft for area calculations.
	COTTER PIN: Eighteen sizes have been standardized in nominal diameters ranging from 1/32″ to 3/4″. Locking device for other fasteners. Used with a castle or slotted nut on bolt, screws, or studs, it provides a convenient, low-cost locknut assembly. Hold standard clevis pins in place. Can be used with or without a plain washer as an artificial shoulder to lock parts in position on shafts.
	CLEVIS PIN: Standard nominal diameters for clevis pins range from 3/16″ to 1″. Basic function of the clevis pin is to connect mating yoke, or fork, and eye members in knuckle-joint assemblies. Held in place by a small cotter pin or other fastening means it provides a mobile joint construction, which can be readily disconnected for adjustment or maintenance.

Fig. 27-2 Machine Pins. (Courtesy Machine Design, Vol. 39, No. 14, 1967)

They employ some form of detent mechanism to provide a locking action in assembly.

Holes for Taper Pins

Holes for taper pins are customarily sized by reaming. A through hole is formed by step drills and straight fluted reamers. The present trend is toward the use of helically

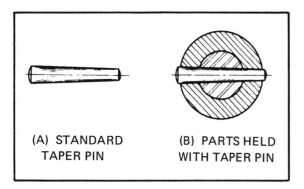

(A) STANDARD TAPER PIN   (B) PARTS HELD WITH TAPER PIN

Fig. 27-3 Taper Pin Application.

fluted taper reamers which provide more accurate sizing and require only a pilot hole the size of the small end of the taper pin. The pin is usually driven into the hole until it is fully seated. The taper of the pin aids hole alignment in assembly.

A tapered hole in a hub and a shaft is shown in figure 27-4. If the hub and shaft are drilled and reamed separately, a misalignment might occur as shown in figure 27-4C. To prevent misalignment, the hub and shaft should be drilled at the same time as the parts are assembled, and each of the detailed parts should carry a note like the following: *DRILL AND REAM FOR NO. 1 TAPER PIN AT ASSEMBLY.*

REFERENCES AND SOURCE MATERIALS

1. F.W. Braendel, "Pin Fasteners," *Machine Design*, 39, (1967).

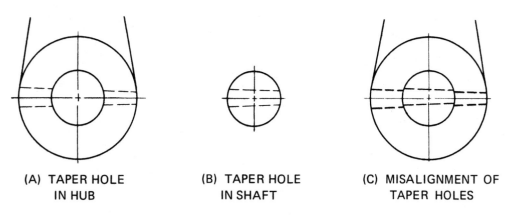

(A) TAPER HOLE
IN HUB

(B) TAPER HOLE
IN SHAFT

(C) MISALIGNMENT OF
TAPER HOLES

Fig. 27-4   Possibility of Hole Misalignment if Holes are Not Drilled at Assembly.

QUESTIONS

1. What views are shown?

2. How many surfaces are to be finished?

3. How many scraped surfaces are indicated?

4. How many holes are to be tapped?

5. What is to fit in tapped hole (R)?

6. What surface is (3) in the left view?

7. What surface is (2) in the left view?

8. What surface is (4) in the front view?

9. What surface is (14) in the left view?

10. What surface in the top view and front view is (9)?

11. What surface in the top view and front view is (8)?

12. What surface is (12) in the top view?

13. What is the name of part (V)?

14. What is the purpose of part (V)?

15. What do dotted lines at (W) represent?

16. What surface is line (6) in the front view?

17. What top view line indicates point (Z)?

18. What is the depth of the tapped hole at (X)?

19. What edges or surfaces in the left and front views does line (T) represent?

20. What is the diameter of tap drill (Y)?

21. Determine dimensions or operations at (A) to (Q)   (20) to (56)

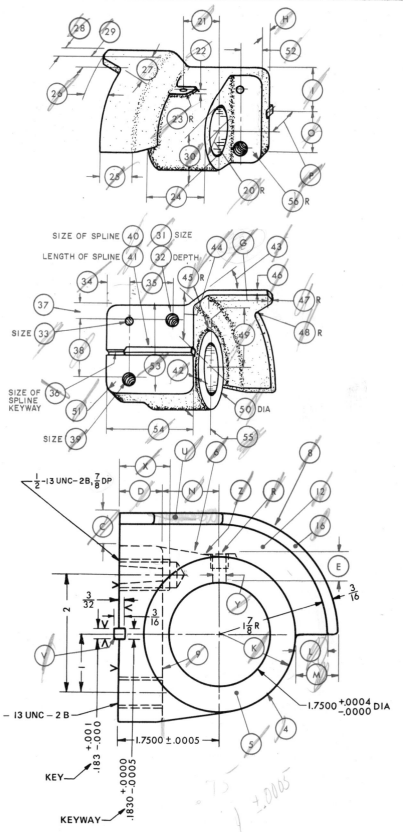

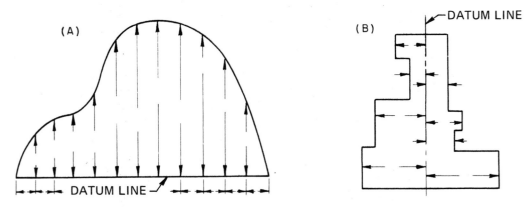

Fig. 28-3 Applications of Datum Dimensioning.

It should be noted that, in this system, the tolerance from the common point to each of the features has to be held to half the tolerance acceptable between individual features. For example, in figure 28-1B, if a tolerance between two individual holes of ± .005 were desired, each of the dimensions shown would have to be held to ± .0025.

REFERENCES AND SOURCE MATERIALS

1. Canadian Standards Association, *Buletin* 78.2 (1967).

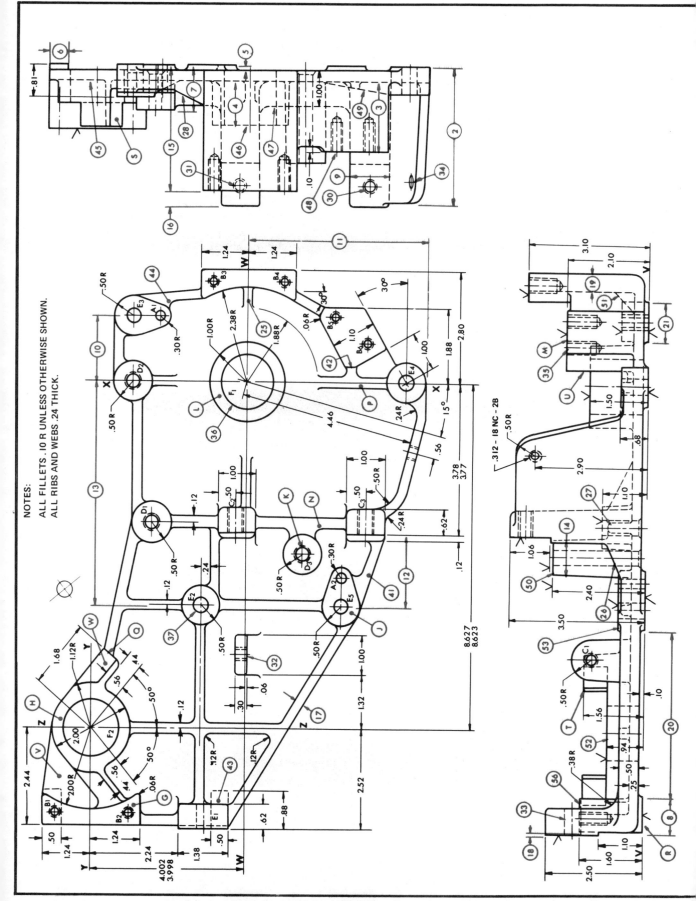

NOTES:
ALL FILLETS .10 R UNLESS OTHERWISE SHOWN.
ALL RIBS AND WEBS .24 THICK.

142

QUESTIONS

NOTE: Where limit dimensions are given use larger limit.

1. What are the overall dimensions of the casting?
2. What symbol indicates a machine surface?
3. Identify surfaces (G) to (U) on one of the other views.
4. Locate rib (25) on the top view.
5. Locate rib (26) on the top view.
6. Locate rib (27) on the side view.
7. Locate rib (28) on the top view.
8. Give the sizes of the following holes: (30) (31) (32) (33) (35) and (34).
9. How deep is hole (35)?
10. What is the tolerance on hole (36)?
11. How deep is hole (37)?
12. Determine distances (2) to (21).

ANSWERS

1 ___
W ___
H ___
2 ___
3 (G) ___ (H) ___ (J) ___ (K) ___ (L) ___ (M) ___ (N) ___ (P) ___ (Q) ___ (R) ___ (S) ___ (T) ___ (U) ___
4 ___
5 ___
6 ___
7 ___
8 (30) ___ (31) ___ (32) ___ (33) ___ (34) ___
9 ___
10 ___
11 ___
12 (2) (3) (4) (5) (6) (7) (8) (9) (10) (11) (12) (13) (14) (15) (16) (17) (18) (19) (20) (21)

HOLE SIZE AND LOCATION

HOLE	DISTANCE FROM					SIZE
	V-V	W-W	X-X	Y-Y	Z-Z	
A₁		2.32	1.62			.252 DIA. / .250 DIA.
A₂		2.50	4.88			
B₁				.94	2.12	
B₂				.94	2.12	.312-18NC-2B x .75 DEEP
B₃		.94	2.50			
B₄		.94	2.50			
B₅		2.28	1.62			
B₆		3.08	.96			
C₁	1.38				1.80	.375-16NC-2B
C₂		3.00	.28			
C₃		3.00	3.12			
D₁		2.50	3.50			.3752 DIA. / .3750 DIA.
D₂		3.00	.00			
D₃		1.48	4.28			CSK .06 x 45°
E₁	1.80			2.94		
E₂		1.18	5.56			.406 DIA.
E₃		3.00	1.62			
E₄		4.12	.00			
E₅		2.50	5.56			
F₁	.00	.00	.00			1.3765 DIA. / 1.3745 DIA.
F₂	.00	.00	.00			

QUANTITY	40
MATERIAL	CI
SCALE	1/2
DRAWN	DATE

INTERLOCK BASE   A-47

COURTESY OF WESTINGHOUSE ELECTRIC AND MANUFACTURING CORPORATION, SOUTH PHILADELPHIA, PENNSYLVANIA
ADAPTED FROM TECHNICAL DRAFTING BY CHARLES H. SCHUMANN, HARPER AND ROW PUBLISHERS, NEW YORK

ASSIGNMENT: DETERMINE DISTANCES OR DIMENSIONS AT THE FOLLOWING:

ANSWERS

#36 (.1065)

A 7/8	H	Q	X	6 6-32 unc-2B 13	20	27
B	J	R	Y	7	14	21	28
C	K	S	Z	8	15	22	29
D	L	T	2	9	16	23	30
E	M	U	3	10	17	24	31
F	N	V	4	11	18	25	32
G	P	W	5	12	19	26	33

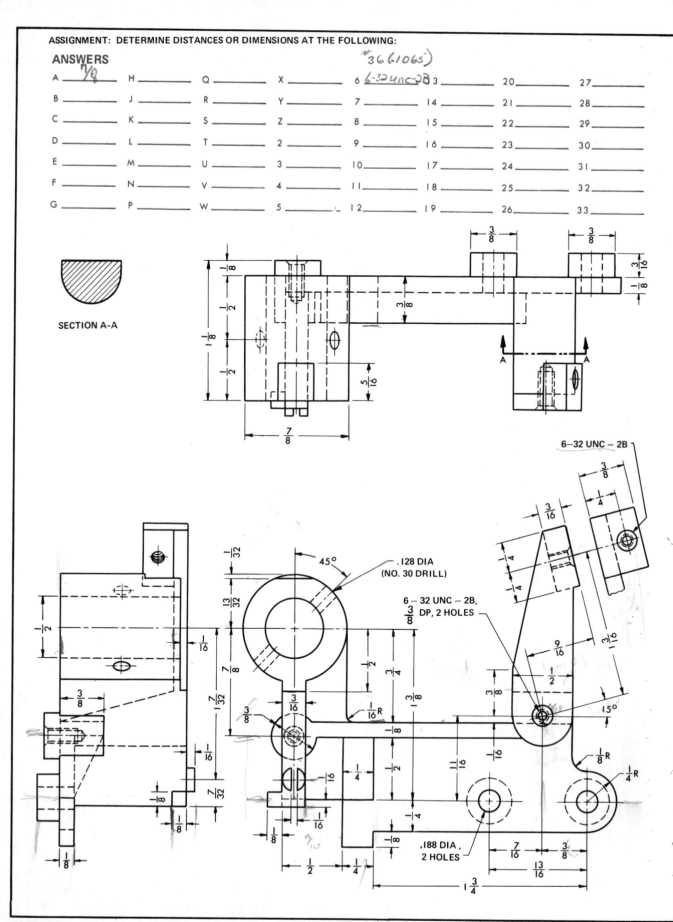

SECTION A-A

6–32 UNC – 2B

.128 DIA (NO. 30 DRILL)

6 – 32 UNC – 2B, 3/8 DP, 2 HOLES

.188 DIA, 2 HOLES

COURTESY OF THE TODD COMPANY, INC. , ROCHESTER, NEW YORK

144

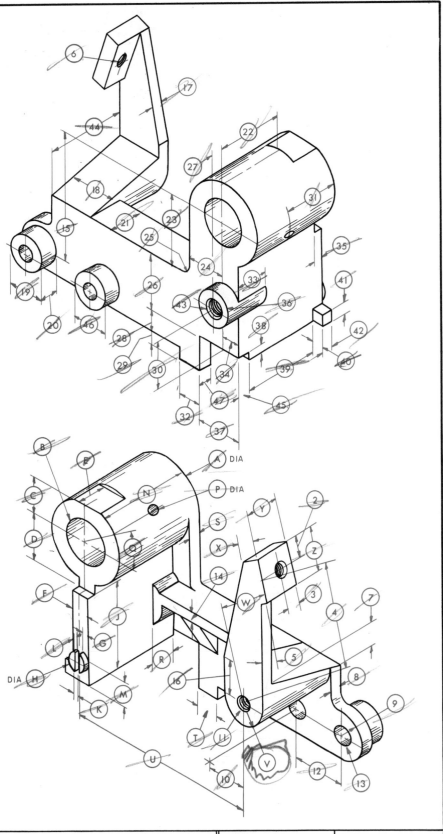

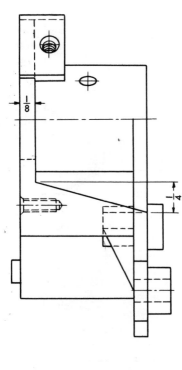

NOTE: TAPPED HOLES TO BE
CSK SLIGHTLY

$\frac{1}{8}$

$\frac{1}{4}$

QUANTITY		DRAWN		DATE
MATERIAL	ALUMINUM	**CONTROL**		
SCALE	2/1	**BRACKET**	**A-48**	

ASSEMBLY DRAWINGS

The term *assembly drawing* refers to that type of drawing in which the various parts of a machine or structure are drawn in their relative positions in the completed unit.

In addition to showing how the parts fit together, the assembly drawing is used mainly:

- To represent the proper working relationships of the mating parts of a machine or structure and the function of each.

- To give a general idea of how the finished product is to look.

- To aid in securing overall dimensions and center distances in assembly.

- To give the detailer sufficient data from which to design the smaller units of a larger assembly.

- To supply illustrations which may be used for catalog or other illustrative purposes.

In order to show the working relationship of interior parts, the principles of projection may be violated and details omitted for clarity. Assembly drawings should not be overloaded with details describing the shape of the parts since information of this nature is given on detail drawings.

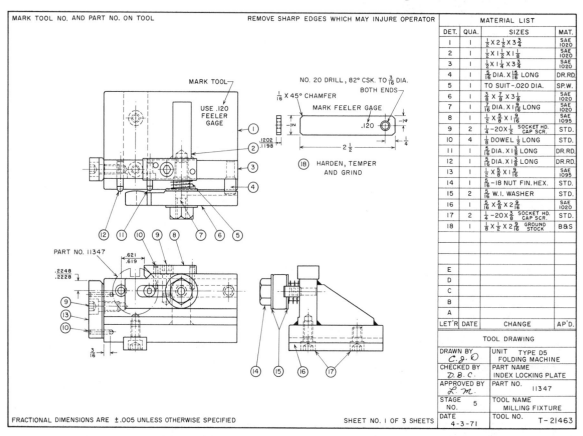

Fig. 29-1   An Assembly Drawing.

Detail dimensions which would tend to confuse the assembly drawing should be omitted. Only such dimensions as center distances, overall dimensions, and dimensions which show the relationship of the parts as they apply to the mechanism as a whole may be included. There are times when a simple assembly drawing may be dimensioned so that no other detail drawings are needed. In such a case the assembly drawing becomes a working assembly drawing.

Sectioning is used more extensively on assembly drawings than on detail drawings. The conventional method of section lining is used on assembly drawings to indicate different materials in the unit and also to show the relationship of the various parts.

Subassembly Drawings

Subassembly drawings are often made of the smaller mechanical units which, when combined in final assembly, make a single machine. In the case of a lathe, subassembly drawings would be furnished for the headstock, the apron, and other units of the carriage. While these units might be machined and assembled in different departments according to the subassembly drawings, the individual units would be combined in final assembly according to the assembly drawing.

Identifying Parts of an Assembly Drawing

When a machine is designed, an assembly drawing or design layout is first drawn to visualize clearly the performance, shape, and clearances of the various parts. From this assembly drawing, the detail drawings are made and each part is given a part number.

To assist in the assembling of the machine, part numbers that correspond with the part numbers of the various details are placed on the assembly drawing. Small circles, .30 inch to .50 inch in diameter, that contain the part number are then attached to the corresponding part with a leader.

SECTION THROUGH SHAFTS, PINS AND KEYS

Shafts, bolts, nuts, rods, rivets, keys, pins, and similar solid parts, the axes of which lie

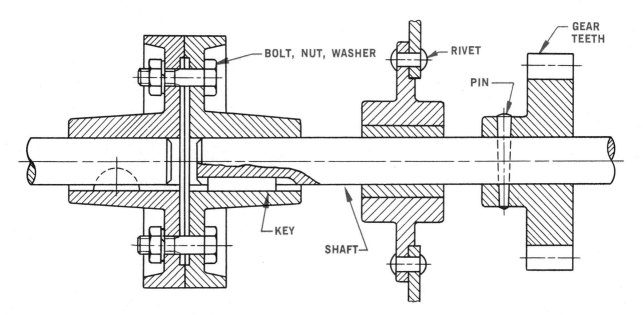

Fig. 29-2 Parts That are Not Section Lined in Section Drawings Even Though the Cutting Line Passes Through.

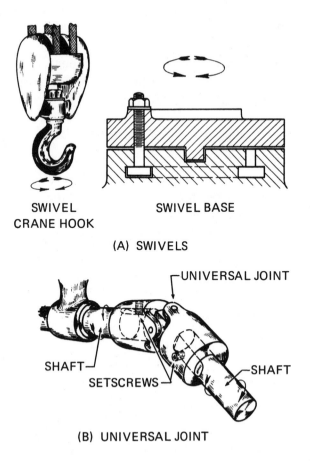

SWIVEL
CRANE HOOK

SWIVEL BASE

(A) SWIVELS

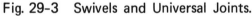

UNIVERSAL JOINT

SHAFT

SETSCREWS

SHAFT

(B) UNIVERSAL JOINT

Fig. 29-3   Swivels and Universal Joints.

on the cutting plane, are not sectioned except when a brokenout section of the shaft is used to indicate clearly the key, keyseat or pin.

SWIVELS AND UNIVERSAL JOINTS

A swivel is composed of two or more pieces which are so constructed that each part rotates in relation to the other about a common axis.

A universal joint is composed of three or more pieces which are designed so as to permit the free rotation of two shafts whose axes deviate from a straight line.

In practice, universal joints are constructed in many different forms, the design of which is influenced by the requirements of the working mechanism and the cost of the parts.

Universal joints are used by the designer where a rotating or swinging motion is desired or where power must be delivered along shafts which are not in a straight line.

COTTER PINS

The cotter pin is a standard machine pin which is commonly used as a fastener in the assembly of such machine parts where no great accuracy is required. While there is no standard way of representing cotter pins in assembly, the manner in which the cotter pin is

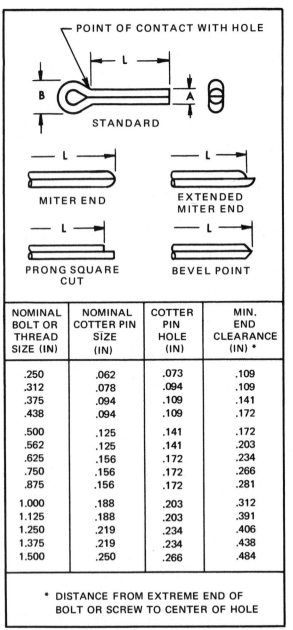

NOMINAL BOLT OR THREAD SIZE (IN)	NOMINAL COTTER PIN SIZE (IN)	COTTER PIN HOLE (IN)	MIN. END CLEARANCE (IN) *
.250	.062	.073	.109
.312	.078	.094	.109
.375	.094	.109	.141
.438	.094	.109	.172
.500	.125	.141	.172
.562	.125	.141	.203
.625	.156	.172	.234
.750	.156	.172	.266
.875	.156	.172	.281
1.000	.188	.203	.312
1.125	.188	.203	.391
1.250	.219	.234	.406
1.375	.219	.234	.438
1.500	.250	.266	.484

* DISTANCE FROM EXTREME END OF BOLT OR SCREW TO CENTER OF HOLE

Fig. 29-4   Cotter Pin Data.

QTY	ITEM	MATL	DESCRIPTION	PT NO
4	NUT-HEX-REG	ST	.375 UNC	7
4	BOLT-HEX-REG	ST	.375 UNC x 1.50 LG	6
I	KEY	MS	WOODRUFF 608	5
2	BEARINGS	SKF	RADIAL BALL 620 OZ	4
I	SHAFT	CRS	1.00 DIA. x 6.50 LG	3
I	SUPPORT	MS	.38 x 2.00 x 4.40	2
I	BASE	CI	PATTERN - A3154	I

Fig. 29-5   Bill of Material.

represented in the assembly of the universal trolley (drawing A-49) is commonly used.

BILL OF MATERIAL

A bill of material is an itemized list of all the components shown on an assembly drawing or a detail drawing. Often, a bill of material is placed on a separate sheet for ease of handling and duplicating. Because the bill of material is used by the purchasing department to order the necessary material for the design, it is necessary to show in the bill of material the size of the raw material rather than the size of the finished part. For castings, a pattern number would appear in the size column in lieu of the physical size of the part.

STANDARD COMPONENTS

Standard components, which are purchased rather than fabricated, such as bolts, nuts and bearings, should have a part number and appear on the bill of material. Sufficient information would be shown in the descriptive column so that the purchasing agent may order these parts.

Standard components are incorporated in the design of machine parts for economical production. These parts are specified on the drawing according to the manufacturer's specification. The use of manufacturers' catalogs is essential for determining detailing standards, characteristics of a special part, methods of representing, etc.

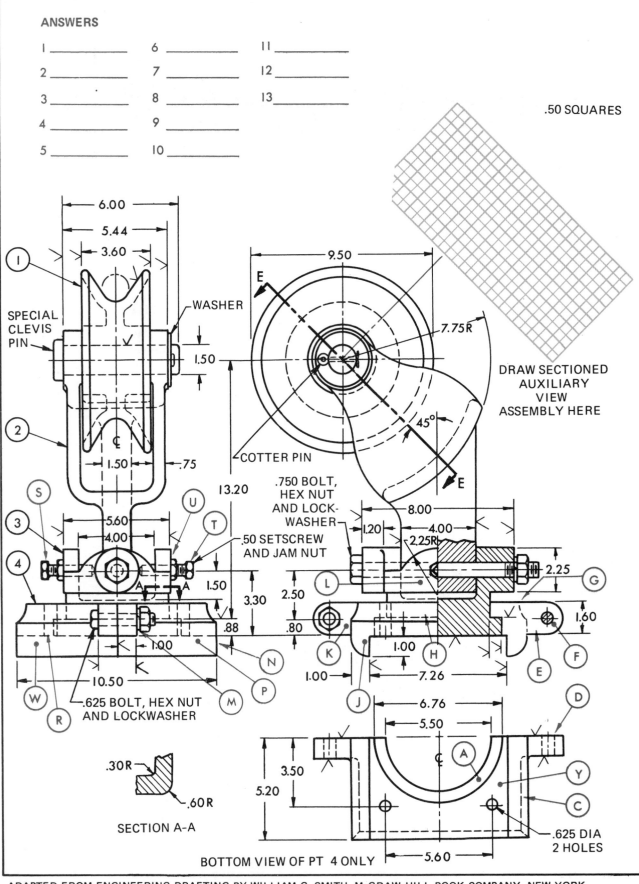

ANSWERS

1 _____ 6 _____ 11 _____
2 _____ 7 _____ 12 _____
3 _____ 8 _____ 13 _____
4 _____ 9 _____
5 _____ 10 _____

.50 SQUARES

6.00
5.44
3.60

I

SPECIAL
CLEVIS
PIN

WASHER

✓

1.50

2

C

1.50 .75

9.50

E

7.75R

DRAW SECTIONED
AUXILIARY
VIEW
ASSEMBLY HERE

45°

E

COTTER PIN

13.20

S

3

U

T

.750 BOLT,
HEX NUT
AND LOCK-
WASHER

.50 SETSCREW
AND JAM NUT

5.60
4.00

4

A   A

1.50

3.30

.88

8.00
1.20
4.00
2.25R

L

2.50

.80

K

1.00

H

2.25

G

1.60

F

E

W

R

1.00

10.50

N

P

M

.625 BOLT, HEX NUT
AND LOCKWASHER

1.00

1.00   7.26

J

6.76

D

.30R

.60R

SECTION A-A

5.50

A

Y

C

3.50

5.20

.625 DIA
2 HOLES

5.60

BOTTOM VIEW OF PT 4 ONLY

ADAPTED FROM ENGINEERING DRAFTING BY WILLIAM G. SMITH, McGRAW–HILL BOOK COMPANY, NEW YORK

SKETCHING ASSIGNMENT

1. Make a sectioned auxiliary view assembly taken on cutting plane line E-E.

2. Make a three-view working drawing of part 3 in the space provided.

BILL OF MATERIAL

Complete the bill of material for the universal trolley assembly and place the part numbers on the assembly drawing. Refer to manufacturer's catalogs and drafting manuals for standard components. Two lines in the bill of material may be used for one part if required. Sizes for cast iron parts need not be shown.

QUESTIONS

1. Locate surface (F) on the bottom view.

2. Is surface (J) shown in the bottom view?

3. What line or surface in the left side view represents surface (J) ?

4. How many different parts are used to make up the trolley?

5. What is the total number of parts used to make up the trolley?

6. How many tapped holes are there excluding the nuts?

7. What is the total number of holes in parts (I) to (4) ? Note: Count through holes as 2 holes.

8. What line or surface in the bottom view represents point (G) ?

9. Locate surface (A) in the front view.

10. Locate surface (Y) in the left side view.

11. What is the overall height of the assembly?

12. Locate surface (L) in the left side view.

13. What is the total number of surfaces that are to be finished for parts (I) to (4) ? Note: One or more surfaces can be on the same plane. Where one finish mark is shown on matched surfaces of castings, this implies that both surfaces are to be finished.

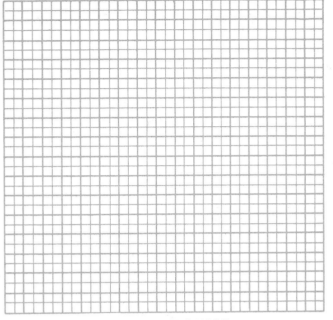

SKETCH PT 3 HERE

QTY	ITEM	MATL	DESCRIPTION	PT NO
	STAND	C I		4
	SWIVEL	C I		3
	SUPPORT	C I		2
	TROLLEY	C I		I

SCALE	1/4

DRAWN	DATE

UNIVERSAL TROLLEY   **A-49**

UNIT

30

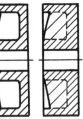

TRUE PROJECTION   PREFERRED METHODS

Fig. 30-1   Conventional Methods of Sectioning Webs.

CONVENTIONAL SECTIONING OF WEBS

The conventional method of representing a section of a part having webs or partitions is preferred to the method of drawing the section in true projection. While the conventional method is a violation of true projection, it is preferred over true projection because of ease in drawing and clearness.

HELICAL SPRINGS

The coil or helical spring is commonly used in machine design and construction. It may be cylindrical or conical in shape or a combination of the two, figure 30-2.

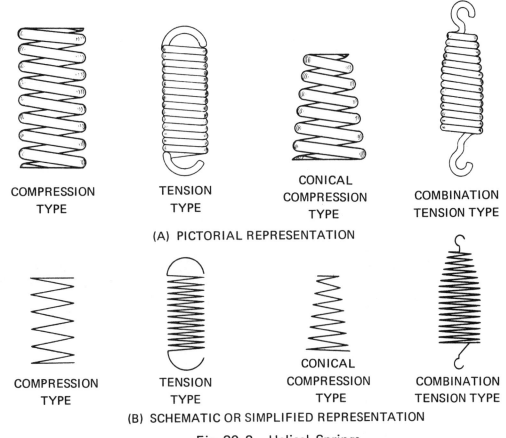

COMPRESSION TYPE   TENSION TYPE   CONICAL COMPRESSION TYPE   COMBINATION TENSION TYPE

(A) PICTORIAL REPRESENTATION

COMPRESSION TYPE   TENSION TYPE   CONICAL COMPRESSION TYPE   COMBINATION TENSION TYPE

(B) SCHEMATIC OR SIMPLIFIED REPRESENTATION

Fig. 30-2   Helical Springs.

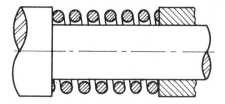

(A) LARGE SPRINGS

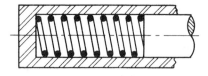

(B) SMALL SPRINGS

Fig. 30-3 Showing Helical Springs on Assembly Drawings.

The true projection of a helical spring is usually not drawn because of the labor and time involved. Instead, a schematic or simplified drawing is preferred because of its simplicity, and all the required information can be given on such a drawing.

On assembly drawings, springs are normally shown in section and either crosshatched lines or solid black· shading is recommended, depending on the size of the wire's diameter.

The following information must be given on a drawing of a spring, figure 30-4:

- Size, shape, and kind of material used in the spring
- Diameter (outside or inside)
- Pitch or number of coils
- Shape of ends
- Length

For example:

Required:

ONE HELICAL TENSION SPRING 3.00 LG (OR NUMBER OF COILS), .50 ID, PITCH .18, 18 B & S GA. SPRING BRASS WIRE

The *pitch* of a coil spring is the distance from the center of one coil to the center of the next. The sizes of spring wires are designated by gage numbers and also in decimal parts of an inch, the tables for which are found in handbooks.

Springs are made to a dimension of either outside diameter (if the spring works in a hole) or inside diameter (if the spring works on a rod). In some cases the mean diameter is specified for computation purposes.

PIPE THREADS

Pipe threads are cut on a taper of .75 inch per foot in order to insure easy starting and provide a tight joint which will hold liquids or gases under high pressures. Pipe threads are cut with standard pipe thread dies and taps.

The standards which have been adopted for pipe threads governing the shape, taper, sizes, pitches, etc., are listed in table form in handbooks.

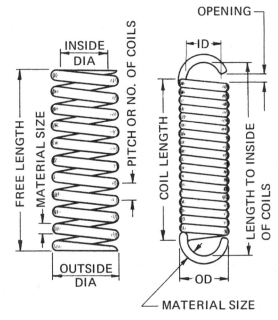

Fig. 30-4 Information Given on Spring Drawings.

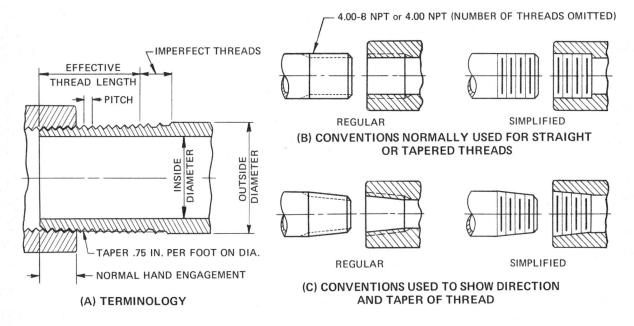

EFFECTIVE
THREAD LENGTH
IMPERFECT THREADS
PITCH
INSIDE DIAMETER
OUTSIDE DIAMETER
TAPER .75 IN. PER FOOT ON DIA.
NORMAL HAND ENGAGEMENT

(A) TERMINOLOGY

4.00-8 NPT or 4.00 NPT (NUMBER OF THREADS OMITTED)

REGULAR   SIMPLIFIED

**(B) CONVENTIONS NORMALLY USED FOR STRAIGHT
OR TAPERED THREADS**

REGULAR   SIMPLIFIED

**(C) CONVENTIONS USED TO SHOW DIRECTION
AND TAPER OF THREAD**

Fig. 30-5   Pipe Thread Terminology and Conventions.

Pipe threads, tapered or straight, are represented by the same conventions as regular screw threads, as shown in figure 30-5. It is not necessary to draw pipe threads with a taper, although a taper of approximately 8 to 1 may be used if desired, but the thread designation must indicate whether straight or taper threads are required.

The notes used to complete the information are of the same general type as those used for screw threads.

Example: 4.00 – 8 NPT means

4.00 – nominal diameter of pipe
8 – number of threads per inch
N – American Standard
P – Pipe
T – Taper thread

The specifications for a 4.00 pipe having a straight thread would read: 4.00 – 8 NPS where S means straight thread. The number of threads may be omitted, since the number of threads per inch on a pipe size does not vary.

SKETCHING ASSIGNMENT

Sketch two views of each of parts 1 and 5 in the space provided on the drawing. Completely dimension the parts.

QUESTIONS

1. How many separate parts are shown on the valve assembly?

2. What is the length of the spring when the valve is closed?

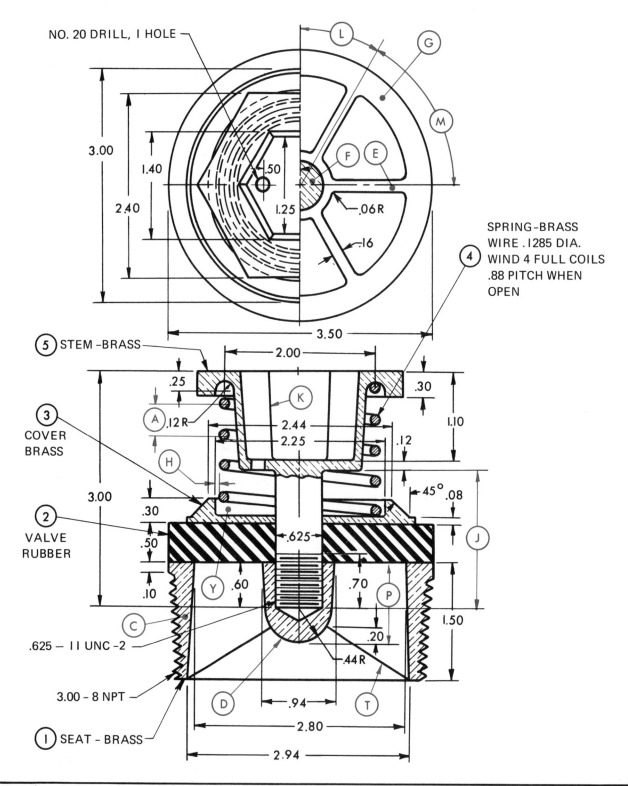

NO. 20 DRILL, I HOLE

3.00

1.40

.50

2.40

1.25

.06R

.16

3.50

SPRING-BRASS WIRE .1285 DIA. WIND 4 FULL COILS .88 PITCH WHEN OPEN

(5) STEM –BRASS

2.00

.25

.30

(3) COVER BRASS

(A) .12R

2.44

2.25

.12

1.10

(H)

45° .08

3.00

.30

.625

(2) VALVE RUBBER

.50

.60

.70

.20

.10

1.50

.625 – II UNC –2

.44R

3.00 – 8 NPT

.94

2.80

(1) SEAT - BRASS

2.94

3. Determine distances Ⓐ Ⓙ Ⓟ .

4. What is the overall free length of spring?

5. Locate Ⓔ in the front view.

6. How many supporting ribs are there connecting Ⓓ to Ⓒ ?

7. How thick are these ribs?

8. Locate parts Ⓕ and Ⓖ in the front view.

9. How many threads per inch are there on the stem (part 5) and what thread designation is used?

10. What is the nominal size of the pipe thread?

11. Give the length of the pipe thread.

12. Determine clearance distance Ⓗ .

13. Determine angles Ⓛ and Ⓜ .

ANSWERS

1 _____	8 F _____
2 _____	G _____
3 A _____	9 _____
J _____	_____
P _____	10 _____
4 _____	11 _____
5 _____	12 _____
6 _____	13 L _____
7 _____	M _____

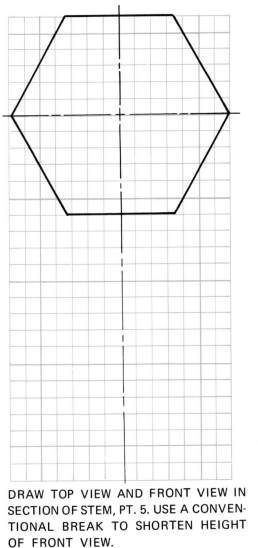

DRAW TOP VIEW AND FRONT VIEW IN SECTION OF STEM, PT. 5. USE A CONVENTIONAL BREAK TO SHORTEN HEIGHT OF FRONT VIEW.

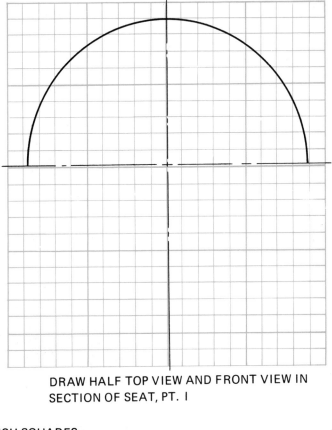

DRAW HALF TOP VIEW AND FRONT VIEW IN SECTION OF SEAT, PT. 1

.20 INCH SQUARES

SCALE	1/1	
DRAWN		DATE
FLUID PRESSURE VALVE		A-50

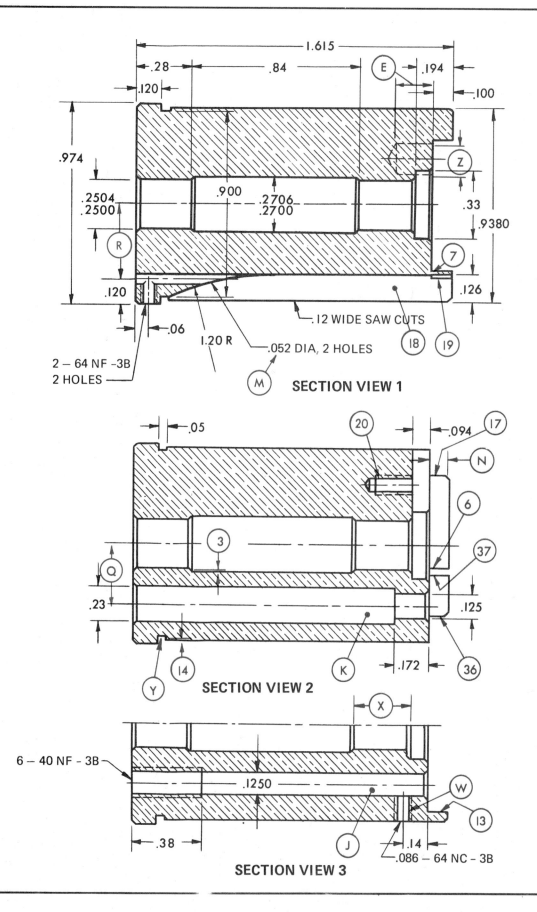

SECTION VIEW 1

SECTION VIEW 2

SECTION VIEW 3

.12 WIDE SAW CUTS

.052 DIA, 2 HOLES

2 – 64 NF –3B
2 HOLES

1.20 R

6 – 40 NF – 3B

.086 – 64 NC – 3B

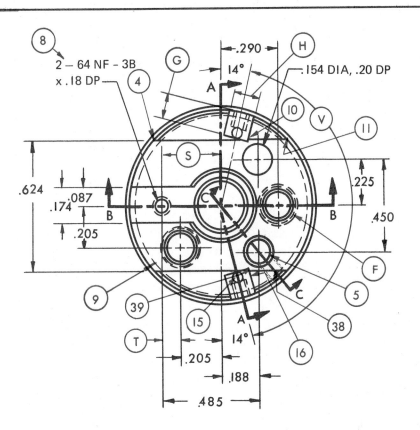

2 – 64 NF – 3B
x .18 DP

.290
14°
.154 DIA, .20 DP

.624
.174 .087
.205
.225
.450
.205
.188
.485
14°

ANSWERS

1 _____
2 _____
3 _____
4 (11) _____
  (19) _____
  (37) _____
  (38) _____
5 _____
6 _____
7 _____
8 _____
9 _____
10 (E) _____
  (G) _____
  (H) _____
  (N) _____
  (Q) _____
  (R) _____
  (S) _____
  (T) _____
  (X) _____

ASSIGNMENT

Label each of the section views with the appropriate
titles with reference to the cutting plane lines on the
end view.

QUESTIONS

1. What is the diameter of hole (F) ?
2. What is the diameter of hole (Z) ?
3. Identify hole (8) in another view.
4. Locate lines (11) (19) (37) (38) in another
   view.
5. Determine angle (V) .
6. Locate line (Y) in another view.
7. Determine depth of slot at (14) .
8. Determine maximum depth of recess at (3) .
9. Locate hole (M) in another view.
10. Determine distances (E) (G) (H) (N) (Q)
    (S) (T) (X)

QTY	6	
MATERIAL	BRASS	
SCALE	2.5/1	
DRAWN		DATE
SPINDLE BEARING	A-51	

UNIT
31

INTERNATIONAL ORGANIZATION FOR STANDARDIZATION [1]

The position of the United States and Canada as two of the important trading nations in the world makes it essential that they participate in the work of *ISO,* the International Organization for Standardization. One section of *ISO* deals with the standardization of drawings for engineering purposes in order to facilitate the preparation, exchange, and reading of drawings.

Because the main type of projection for engineering drawings is orthographic, and because orthographic drawings may be drawn in either third-angle projection (used in Canada and the United States) or first-angle projection (used principally in Europe) it is desirable to indicate the method of projection on drawings which may be used overseas.

ISO recommends the use of the projection symbol shown in figure 31-1 as a means of identifying the type of projection. The symbol should be located in the lower right-hand corner of the drawing adjacent to the title block.

FIRST ANGLE   THIRD ANGLE

Fig. 31-1   ISO Projection Symbols.

METRIC DIMENSIONING [1]

With the growth of transportation came the exchange of products between Europe and America. The exchange of drawings and the need to understand the metric system of dimensioning accompanied transatlantic trade.

Dimensions in the metric system are expressed in millimeters except for installation and floor plan drawings which are usually expressed in meters.

All metric dimensions are followed by the abbreviation *MM* for millimeters, or *M* for meters, except when a drawing is completely in the metric system and carries a note that all dimensions are in millimeters or meters.

When dimensions in both the inch and metric systems are used on the same drawing, the controlling dimension, i.e., the dimension in which the product was designed, is given above the dimension line, and the converted value in brackets below the line, figure 31-2.

When direct tolerances or limits of size are required, they are shown with the maximum limit always preceding the minimum limit.

ARRANGEMENT OF VIEWS

Parts which are to be fitted over shafts as a single unit are sometimes made in two or

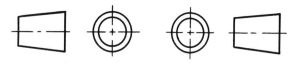

1.80

(45.72 MM)

35 MM

1.378"

(A) DUAL DIMENSIONING

1.82/1.78

(46.22/45.22 MM)

1.80 $^+_-$ .02

(45.72 $^+_-$ .50 MM)

(B) DUAL TOLERANCING

Fig. 31-2   Combined Inch and Metric Systems of Dimensioning.

more pieces. This practice is followed for ease in assembling and replacing on the main structure of a machine rather than for ease in manufacture. Drawing A-52 shows two parts which are bolted and doweled together to form one unit.

The arrangement of views of the spider is illustrated by the diagrams in figure 31-3. By comparing this figure with the drawing of the spider, note that the two halves together represent the top view. The right view is a full section of each half (**A** and **B**). The front view is a drawing of the front of part **A** only.

Although the front and side views are incomplete, the manner in which they are drawn and the arrangement of the views satisfies the demand for clearness and for economy in time and space.

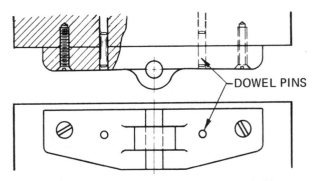

Fig. 31-4   Aligning Parts With Dowel Pins.

DOWEL PINS

Dowel pins or small cylindrical pins are used for a variety of purposes in holding parts in alignment and in guiding parts in some desired position. The most common application of dowel pins is in the aligning of parts which are fastened with screws or bolts and must be accurately fitted together.

When two pieces are to be fitted together as in the case of the part in figure 31-4, one method of aligning is to clamp the two pieces in the desired location, drill and ream the dowel holes, insert the dowel pins, and then drill and tap for the screw holes.

Drill jigs are frequently used when production in the interchangeability of parts is required or when the nature of the piece itself does not permit the transfer of the doweled holes from one piece to the other. A drill jig was used in drilling the dowel holes for the spider (drawing A-52).

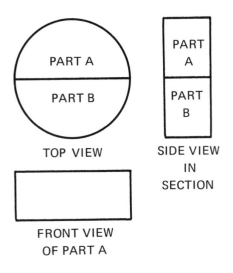

Fig. 31-3   Arrangement of Views for Spider, Drawing A-52.

REFERENCES AND SOURCE MATERIALS

1. Canadian Standards Association, *Bulletin* 78.2 (1967).

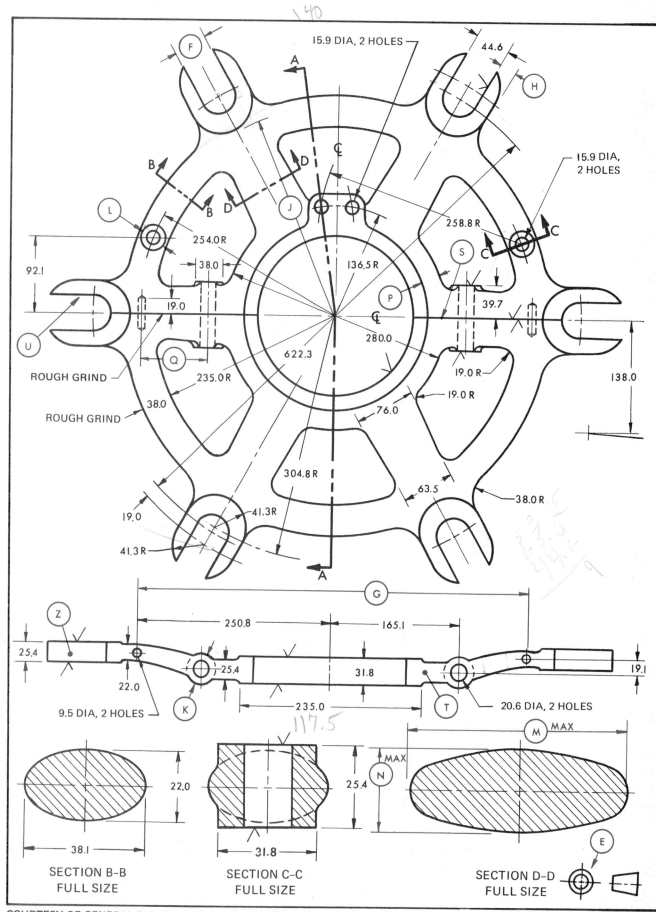

15.9 DIA, 2 HOLES

15.9 DIA, 2 HOLES

44.6

258.8 R

254.0 R

38.0

19.0

136.5 R

92.1

280.0

ROUGH GRIND

19.0 R

235.0 R

622.3

39.7

19.0 R

19.0 R

138.0

ROUGH GRIND

38.0

76.0

304.8 R

63.5

38.0 R

19.0

41.3 R

41.3 R

G

250.8

165.1

Z

25.4

22.0

25.4

31.8

19.1

9.5 DIA, 2 HOLES

K

235.0

T

20.6 DIA, 2 HOLES

M MAX

N MAX

22.0

25.4

38.1

31.8

SECTION B-B
FULL SIZE

SECTION C-C
FULL SIZE

SECTION D-D
FULL SIZE

E

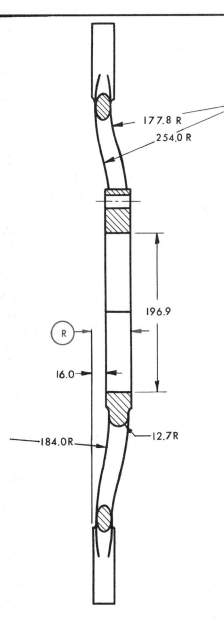

SECTION A-A

ASSIGNMENT

Below or beside each of the dimensions shown, place the equivalent value in inches correct to two decimal places. Refer to the tables in the back of the text.

QUESTIONS

1. What is the unit of measurement on this drawing?

2. What type of projection is indicated by the symbol E ?

3. How many different size scales are used on the drawing?

4. Locate surface T in the top view.

5. Locate surface Z in the top view.

6. What is the approximate extreme outside diameter of the spider?

7. What will be the rough dimension of casting at F, assuming that 1.5 mm. has been added for each surface to be finished?

8. How many different surfaces are to be finished on the lower half of the spider?

9. How many different surfaces are to be finished on the upper half of the spider?

10. Calculate distances G H J K L M N P Q R .

ANSWERS

1. _Decimal_
2. _____
3. _____
4. _____
5. _____
6. _____
7. _____
8. _____
9. _____
10. G _____
    H _____
    J _____
    K _____
    L _____
    M _____
    N _____
    P _____
    Q _____
    R _____

NOTE: DIMENSIONS SYMMETRICAL ABOUT ₵

NOTE: ALL DIMENSIONS IN MILLIMETERS

QUANTITY	300	
MATERIAL	CAST IRON	
SCALE	1/4 EXCEPT WHERE NOTED	
DRAWN		DATE
SPIDER		A-52

UNIT
32

PARTIAL VIEWS

A partial view is one in which only part of an object is shown. However, sufficient information is given in the partial view to complete the description of the object. The partial view is limited by a break line. Partial views are used because of the following advantages:

- They save time in drawing.

- They conserve space which might otherwise be required to draw the object.

- They sometimes permit the drawing to be made to a scale large enough to bring out all details clearly, whereas if the whole view were drawn, lack of space might make it necessary to draw to a smaller scale, and as a result the clearness of some of the details might be lost.

- If the part is symmetrical, a partial view (referred to as a half or quarter view) may be drawn on one side of the centerline as shown in figure 32-2. In the case of the coil frame (drawing A-53) a partial view was used so that the object could be drawn to a larger scale for clearness, thus saving time and space.

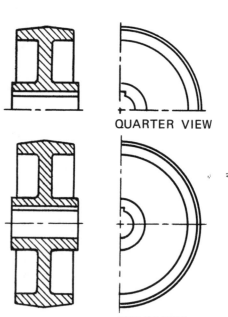

QUARTER VIEW

HALF VIEW

Fig. 32–2 Half and Quarter Views.

RIBS IN SECTION

A true projection section view (figure 32-3A), would be misleading when the cutting plane passes longitudinally through the center of the rib. To avoid this impression of solidity, a preferred section not showing the ribs section-lined or crosshatched is used. When there is an odd number of ribs (figure 32-3B), the top rib is aligned with the bottom rib to show its true relationship with the hub and flange. If the rib is not aligned or revolved, it appears distorted on the section view and is misleading.

An alternative method of identifying ribs in a section view is shown in figure 32-3C, a base in section. If rib **A** of the base were not sectioned as previously mentioned, it would appear exactly like **B** in the section view and would be misleading. To distinguish between the ribs on the base, alternate section lining on the ribs is used. The line between the rib and solid portions is shown as a broken line.

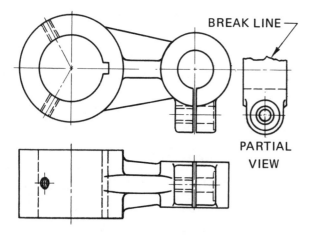

BREAK LINE

PARTIAL VIEW

Fig. 32–1   Partial Views.

HOLES ARE ROTATED TO CUTTING PLANE TO SHOW THEIR
TRUE RELATIONSHIP WITH THE REST OF THE ELEMENT

RIBS ARE NOT SECTIONED

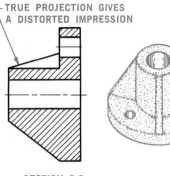

SECTION A-A
PREFERRED

SECTION A-A
TRUE PROJECTION

(A) CUTTING PLANE PASSING THROUGH BOTH RIBS

TRUE PROJECTION GIVES
A DISTORTED IMPRESSION

SECTION B-B
PREFERRED

SECTION B-B
TRUE PROJECTION

HOLE AND RIB ARE ROTATED
TO CUTTING PLANE

(B) CUTTING PLANE PASSING THROUGH ONE RIB AND ONE HOLE

RIB B

RIB A

RIB B

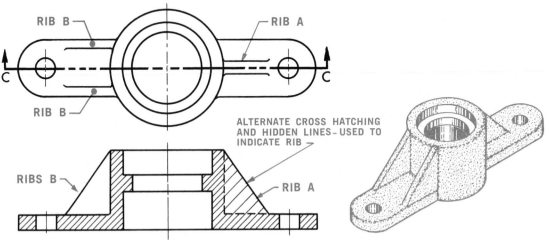

ALTERNATE CROSS HATCHING
AND HIDDEN LINES – USED TO
INDICATE RIB

RIBS B

RIB A

SECTION C-C

(C) ALTERNATE METHOD OF SHOWING RIBS IN SECTION

Fig. 32-3   Ribs in Section.

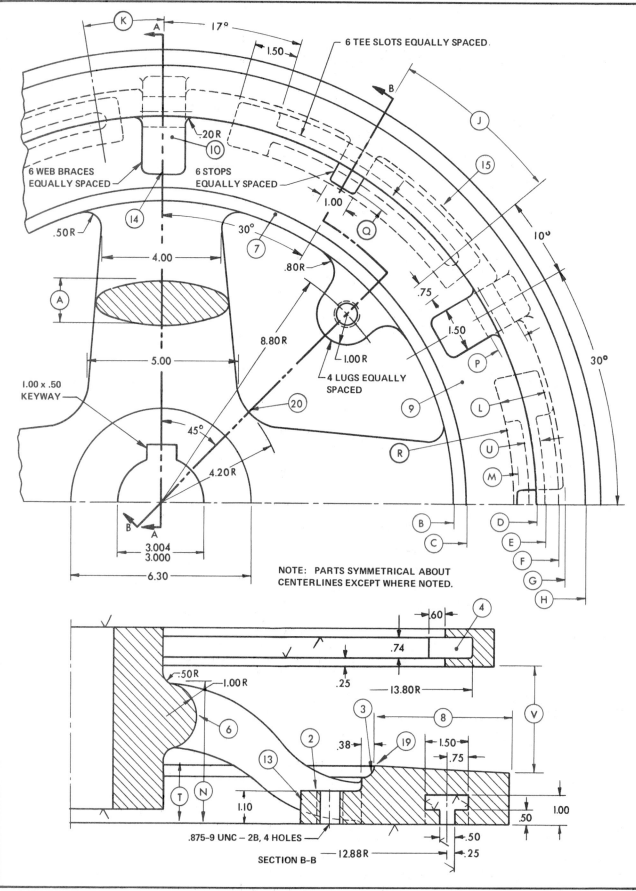

6 TEE SLOTS EQUALLY SPACED.

17°

1.50

6 WEB BRACES EQUALLY SPACED

6 STOPS EQUALLY SPACED

.20R

.50R

4.00

30°

.80R

1.00

8.80R

1.00R

5.00

4 LUGS EQUALLY SPACED

1.00 x .50 KEYWAY

45°

4.20R

3.004
3.000

6.30

.75

1.50

30°

NOTE: PARTS SYMMETRICAL ABOUT CENTERLINES EXCEPT WHERE NOTED.

.60

.74

.25

13.80R

.50R

1.00R

.38

1.50

.75

1.10

1.00

.50

.875-9 UNC – 2B, 4 HOLES

SECTION B-B

12.88R

.50

.25

166

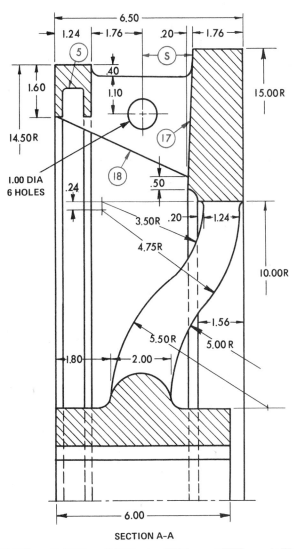

SECTION A-A

QUESTIONS

1. What is the name of part ② ?
2. What is the distance Ⓐ , assuming the revolved section was taken at the middle of the arm?
3. What is the total quantity of part ④ ?
4. Locate surface ⑤ on the plan view.
5. Locate point ⑥ on the plan view.
6. How far is line ⑭ from the center point?
7. How far is surface ⑬ from the center point?
8. What point on section B-B represents radius Ⓒ ?
9. What line on section B-B represents surface ⑦ ?
10. What is the radius of surface ③ ?
11. Determine distance ⑧
12. What line on section A-A represents surface ⑩ ?
13. What surface on the plan view represents line ⑰ ?
14. What is the angle at Ⓙ ?
15. What is the angle at Ⓚ ?
16. What is the thickness of the web brace?
17. What is the name of part ④ ?
18. What is the thickness of the lugs?
19. What is the distance from the 1.00 dia. hole to the center of the coil frame?
20. Determine distances Ⓛ Ⓝ Ⓟ Ⓠ Ⓢ Ⓣ Ⓤ Ⓥ
21. Determine radii Ⓑ Ⓒ Ⓓ Ⓔ Ⓕ Ⓖ Ⓗ Ⓜ Ⓡ

ANSWERS

1. _____
2. _____
3. _____
4. _____
5. _____
6. _____
7. _____
8. _____
9. _____
10. _____
11. _____
12. _____
13. _____
14. _____
15. _____
16. _____
17. _____
18. _____
19. _____
20. Ⓛ _____
    Ⓝ _____
    Ⓟ _____
    Ⓠ _____
    Ⓢ _____
    Ⓣ _____
    Ⓤ _____
    Ⓥ _____
21. Ⓑ _____
    Ⓒ _____
    Ⓓ _____
    Ⓔ _____
    Ⓕ _____
    Ⓖ _____
    Ⓗ _____
    Ⓜ _____
    Ⓡ _____

QUANTITY	400	
MATERIAL	CAST IRON	
SCALE	1/2	
DRAWN		DATE

COIL FRAME   **A-53**

167

STRUCTURAL STEEL SHAPES[1]

Structural steel is widely used in the metal trades for the fabrication of machine parts because the variety of standard shapes lend themselves to many different types of construction.

It is important to remember that the steel produced at the rolling mills and shipped to the fabricating shop comes in a wide variety of shapes (approximately 600) and forms. At this stage it is called plain material.

The great bulk of this material, as shown in figure 33-1, can be designated as one of the following:

- Standard beams (generally called I-beams because of their resemblance to the capital letter I). These shapes are rolled in many sizes (3 in. to 20 in.).

- Standard channels (3 in. to 18 in.).

- Wide flange (WF) (6 in. to 36 in.) and welded wide flange (WWF) (27 in. to 48 in.) beams and column, sometimes referred to as H-shapes.

- Joists and light beams which are lightweight shapes similar in contour to WF beams (6 in. to 16 in.).

- Structural tees made by slitting the webs of standard I-beams, wide flange beams, joists, and junior beams, thus forming two T-shapes from each beam.

- Angles, consisting of two legs, set at right angles (.75 in. to 8 in.).

- Plates and round and rectangular bars, which are rolled by most producers.

Abbreviations

When rolled steel shapes are designated on drawings, it is desirable that a standard method of abbreviating be followed that will identify the group of shapes without reference to the manufacturer and without the use of in. and lb. marks.

To this end, it is recommended that the nominal depth of the shape, its group symbol, and its weight in pounds per linear foot be abbreviated as illustrated in figure 33-2.

The abbreviations shown are intended only for use on design drawings. When lists of materials are being prepared for ordering from the mills, the requirements of the respective mills from which the material is to be ordered should be observed.

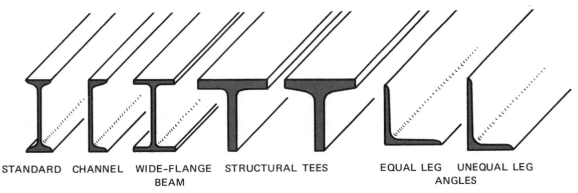

STANDARD   CHANNEL   WIDE-FLANGE   STRUCTURAL TEES   EQUAL LEG   UNEQUAL LEG
BEAM   ANGLES

Fig. 33-1   Common Structural Steel Shapes.

SHAPE	EXAMPLE
WIDE FLANGE SHAPES	24 WF 76
MISCELLANEOUS LIGHT BEAMS	6 B 12
MISCELLANEOUS LIGHT COLUMNS	6 M 20
JUNIOR BEAMS	7 JR. 5.5
JUNIOR CHANNELS	I0 JR. ⌐ 8.4
STANDARD BEAMS	15 I 42.9
STANDARD CHANNELS	9 ⌐ I3.4
STRUCTURAL TEES	
CUT FROM WF SHAPES	ST 5 WF I0.5
CUT FROM STANDARD BEAMS	ST 6 I 20.4
CUT FROM LIGHT BEAMS	ST 6 B 9.5
CUT FROM JUNIOR BEAMS	ST 6 JR. 5.9
BEARING PILES	I0 BP 42
EQUAL LEG ANGLES	∟ 3 x 3 x .25
(LEG DIMENSIONS x THICKNESS, ALL IN INCHES)	
UNEQUAL LEG ANGLES	∟ 6 x 4 x .50
(LEG DIMENSIONS x THICKNESS, ALL IN INCHES)	

Fig. 33–2   Abbreviations for Structural Steel Shapes.

All standard beams and channel shapes have a slope on the inside flange of 16.67%. (16.67% slope is equivalent to 9° 18′ or a bevel of 2 in. in 12 in.) The miscellaneous wide-flange beam and column shapes have parallel face flanges.

PHANTOM OUTLINES

When it is desired to show parts or mechanisms that are not included in the actual detail or assembly drawing, but which, if shown, will make clear how the mechanism will connect with or operate from an adjacent part, this other part is shown by means of light dash lines (one long line and two short dashes) in the operating position. Such a drawing of the extra part is known as a phantom drawing or view drawn *in phantom.*

On the drawing of the four-wheel trolley (drawing A-54), the track upon which the wheels run is an I-beam. The wheels are set at an angle to the vertical plane in order to ride upon the sloping bottom flange of the I-beam.

The outline of the beam is shown by dash lines and, while not an integral part of the trolley, the outline or phantom view of the I-beam shows clearly how the trolley operates.

The four-wheel trolley (drawing A-54) includes many standard parts in the completed assembly. For example, grease cups, lock washers, Hyatt roller bearings, rivets, and nuts, all of which are standard purchased items, are used. These parts are not detailed but are described in the bill of material. On the other hand, the special countersunk head bolts and the taper washers, commonly called *Dutchmen,* are not standard parts and must therefore be made especially for this particular assembly.

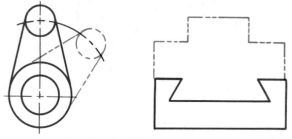

Fig. 33–3   Phantom Lines.

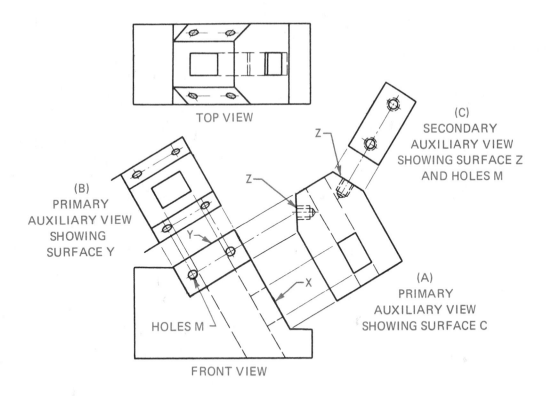

Fig. 33-4   Primary and Secondary Auxiliary Views.

PRIMARY AND SECONDARY AUXILIARY VIEWS

As mentioned earlier, auxiliary views are used to show the true lengths of lines and the true shapes of surfaces which cannot be described in the ordinary views.

A primary auxiliary view is drawn by projecting lines from a regular view where the inclined surface appears as an edge.

The auxiliary view in figure 33-4A, which shows the true projection and true shape of surface **X** is called a *primary auxiliary view* because it is projected directly from the regular front view.

The chamfered corners **Z** have threaded holes on the chamfered surfaces. To show the true shape and location of the threaded holes **M** a second auxiliary view must be shown, as at (C). If this auxiliary view is projected from the first or primary auxiliary view, it is known as a *secondary auxiliary view*. The view at (B) is a primary auxiliary view because it is projected from one of the regular views.

The four-wheel trolley (drawing A-54) is typical of the method of representing only the views necessary for a complete description of the mechanism. The auxiliary view, which takes the place of a front view, is projected at right angles to face **M–M** and shows that part of the cross section view to the right of the centerline **N–N**. The complete cross section view on the drawing of the trolley takes the place of the regular left view.

Note that the projections of parts (5) and (S) shown in the auxiliary view at (K) and (R) are shown as rotated projection drawings.

REFERENCES AND SOURCE MATERIALS

1. Canadian Institute of Steel Construction.

SKETCHING ASSIGNMENT

1. Sketch the secondary auxiliary view of part (A) in the space provided.

2. In the space provided, make a detail drawing complete with dimensions of parts (C) and (D)

QUESTIONS

1. What does hidden line (E) indicate?

2. Which cutting plane in the primary auxiliary view indicates (A) where the section to the left of line N–N is taken, (B) where the section to the right of line N–N is taken?

3. What is the slope of angle (J) ?

4. Locate part (K) in the section view.

5. What is the wheel diameter of the trolley?

6. Locate parts (2)(3)(4)(5)(A)(V)(Z) in the primary auxiliary view.

7. What are the names of parts (T) (U) (V) (W) (X) (Y) ?

8. What is the diameter of the bearing rollers?

9. Determine distance (L)

ANSWERS

1 _____

2 A _____

   B _____

3 _____

4 _____

5 _____

6 (2) _____

(3) _____

(4) _____

(5) _____

(A) _____

(V) _____

(Z) _____

7 (T) _____

(U) _____

(V) _____

(W) _____

(X) _____

(Y) _____

8 _____

9 _____

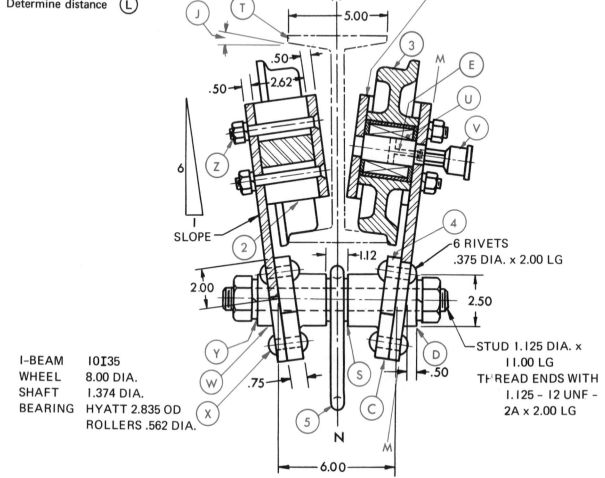

I-BEAM   10I35
WHEEL   8.00 DIA.
SHAFT   1.374 DIA.
BEARING   HYATT 2.835 OD
           ROLLERS .562 DIA.

6 RIVETS
.375 DIA. x 2.00 LG

STUD 1.125 DIA. x
11.00 LG
THREAD ENDS WITH
1.125 – 12 UNF –
2A x 2.00 LG

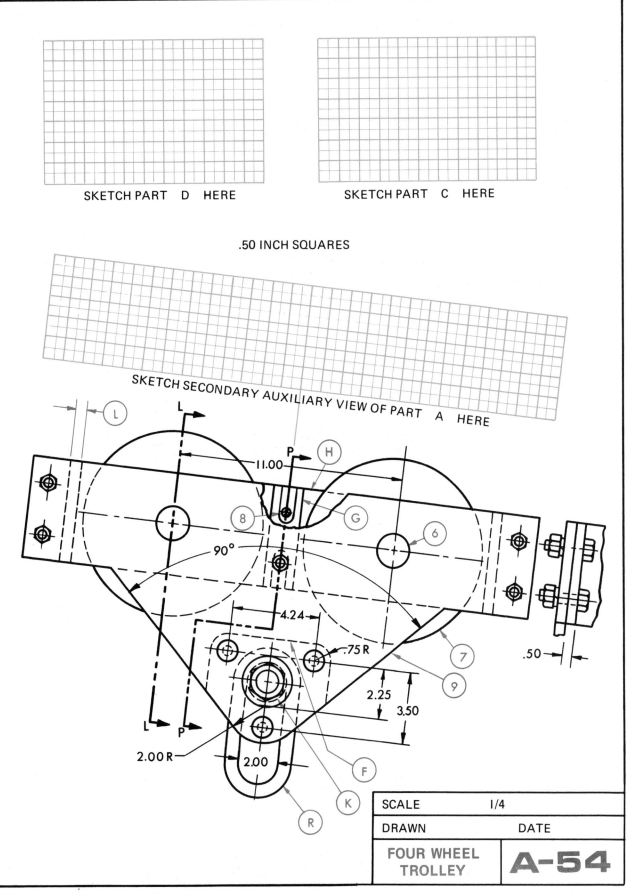

SKETCH PART D HERE

SKETCH PART C HERE

.50 INCH SQUARES

SKETCH SECONDARY AUXILIARY VIEW OF PART A HERE

11.00

90°

4.24

.75 R

2.25

3.50

2.00 R

2.00

.50

SCALE	1/4
DRAWN	DATE

FOUR WHEEL
TROLLEY

A-54

173

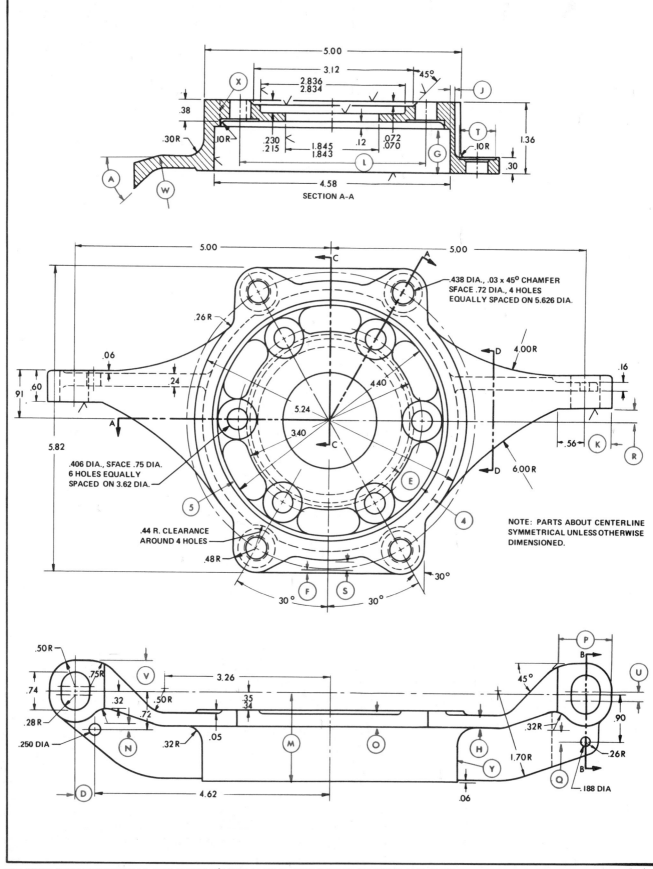

SECTION A-A

.438 DIA., .03 x 45° CHAMFER
SFACE .72 DIA., 4 HOLES
EQUALLY SPACED ON 5.626 DIA.

.406 DIA., SFACE .75 DIA.
6 HOLES EQUALLY
SPACED ON 3.62 DIA.

.44 R. CLEARANCE
AROUND 4 HOLES

NOTE: PARTS ABOUT CENTERLINE
SYMMETRICAL UNLESS OTHERWISE
DIMENSIONED.

.250 DIA

.188 DIA

SECTION B-B
.10 SQUARES

SECTION C-C
.10 SQUARES

SECTION D-D
.10 SQUARES

SKETCHING ASSIGNMENT

In the spaces provided, draw sections B–B, C–C and D–D.

QUESTIONS

1. How many through holes are shown?

2. How many surfaces are to be finished?

3. Determine angle (A) .

4. Determine radius (W) .

5. Locate (4) in the front view.

6. What line in section A–A indicates the surface represented by line (5) ?

7. Determine distances (D) to (V) .

ANSWERS

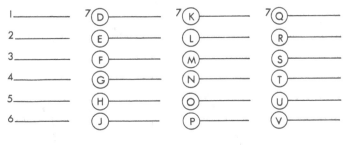

1 _____	7 (D) _____	7 (K) _____
2 _____	(E) _____	(L) _____
3 _____	(F) _____	(M) _____
4 _____	(G) _____	(N) _____
5 _____	(H) _____	(O) _____
6 _____	(J) _____	(P) _____

7 (Q) _____
(R) _____
(S) _____
(T) _____
(U) _____
(V) _____

QUANTITY	400	
MATERIAL	C I	
SCALE	I/I	
DRAWN		DATE

CASE COVER **A-55**

175

UNIT
34

MACHINED LUGS

The design and nature of certain parts make them difficult to be held and machined without the use of lugs. The lugs are an integral part of the casting which are sometimes removed so as not to interfere with the working of the part. Lugs are usually represented in phantom outline.

In each corner of the top and front views of the drive housing (drawing A-56) rectangles are outlined with phantom lines to represent the lugs which are molded on the casting. They are used to keep the part in position during machining operations and are later machined off.

Another example of a machining lug is shown in figure 34-1. This lug is used to provide a flat surface on the end of the casting for centering and also to give a uniform center

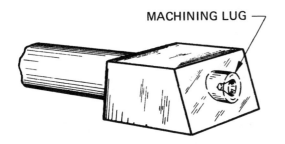

MACHINING LUG

Fig. 34-1   Application of Machining Lug.

bearing for other machining operations. How the part is to be used and its appearance determine whether or not the lug is removed after machining.

FINISHES

Machined parts are frequently finished to protect the surfaces from oxidation or for appearance. The type of finish depends on the way in which a part is to be used. The finishes commonly applied may be either a protective coat of paint, lacquer, or a metallic plating. In some cases only a surface finish such as polishing or buffing may be specified.

The finish may be applied before any machining is done, between the various stages of machining, or after the piece has been completed. The information concerning the type of finish is usually specified on the drawing in a notation similar to the one indicated on the drive housing (drawing A-56) which reads *CASTING TO BE PAINTED WITH ALUMINUM BEFORE MACHINING.*

The decorative finish on the drive housing is to be added before machining so that there will be no accidental deposit of paint on the machined surfaces. This will insure the desired accuracy when assembled with other parts.

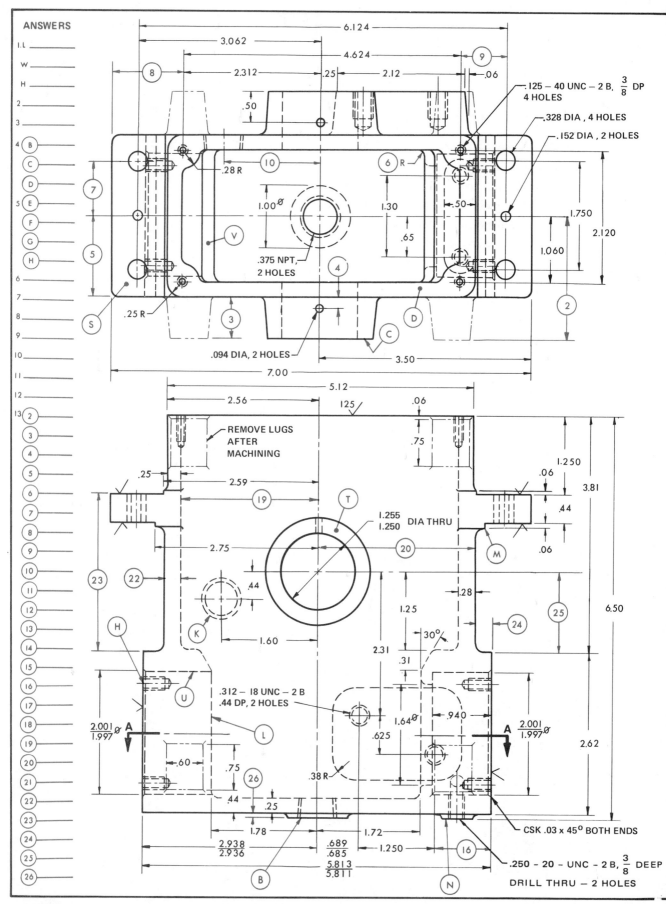

.125 – 40 UNC – 2 B, $\frac{3}{8}$ DP
4 HOLES

.328 DIA , 4 HOLES

.152 DIA , 2 HOLES

6.124
3.062
4.624
9
2.312
.25
2.12
.06
8
.50
10
6   R
1.00 Ø
1.30
.50
1.750
V
.65
2.120
1.060
5
.375 NPT,
2 HOLES
4
7
.28 R
.25 R
S
3
2
.094 DIA, 2 HOLES
C
D
3.50
7.00

5.12
2.56
125
.06
REMOVE LUGS
AFTER
MACHINING
.75
.25
1.250
2.59
.06
19
3.81
T
.44
1.255 DIA THRU
1.250
.06
2.75
20
M
23   22
30°
.28
25
6.50
K
1.60
24
2.31
H
1.25
U
.31
.312 – 18 UNC – 2 B
.44 DP, 2 HOLES
L
1.64 Ø
.940
.625
2.001 Ø
1.997
A
2.001 Ø
1.997
A
2.62
.60
.75
26
.38 R
.44
.25
CSK .03 x 45° BOTH ENDS
1.78
1.72
2.938
2.936
.689
.685
1.250
16
.250 – 20 – UNC – 2 B, $\frac{3}{8}$ DEEP
5.813
5.811
B
N
DRILL THRU – 2 HOLES

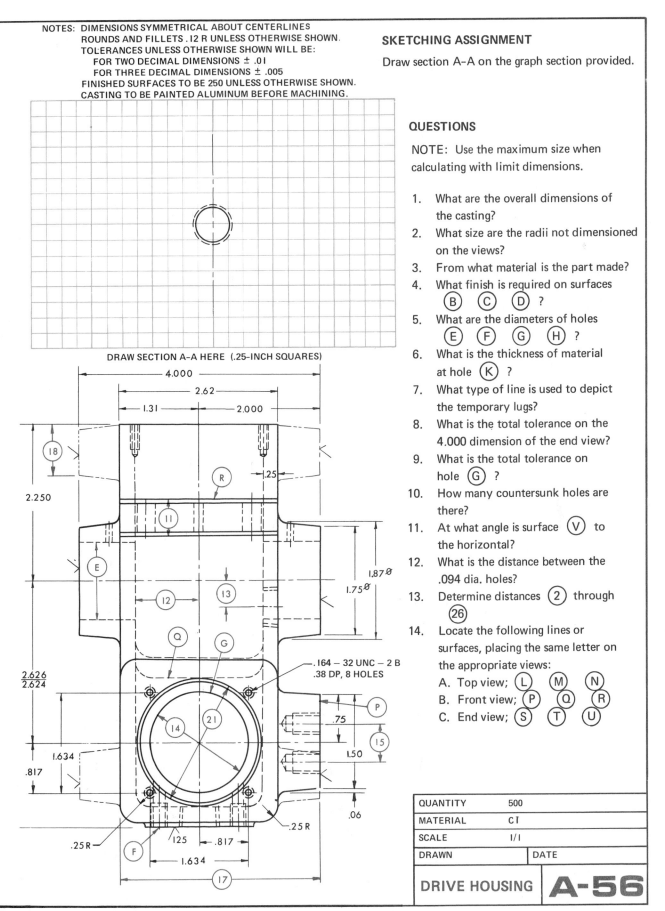

NOTES: DIMENSIONS SYMMETRICAL ABOUT CENTERLINES
ROUNDS AND FILLETS .12 R UNLESS OTHERWISE SHOWN.
TOLERANCES UNLESS OTHERWISE SHOWN WILL BE:
FOR TWO DECIMAL DIMENSIONS ± .01
FOR THREE DECIMAL DIMENSIONS ± .005
FINISHED SURFACES TO BE 250 UNLESS OTHERWISE SHOWN.
CASTING TO BE PAINTED ALUMINUM BEFORE MACHINING.

SKETCHING ASSIGNMENT

Draw section A-A on the graph section provided.

DRAW SECTION A-A HERE (.25-INCH SQUARES)

QUESTIONS

NOTE: Use the maximum size when calculating with limit dimensions.

1. What are the overall dimensions of the casting?
2. What size are the radii not dimensioned on the views?
3. From what material is the part made?
4. What finish is required on surfaces (B) (C) (D) ?
5. What are the diameters of holes (E) (F) (G) (H) ?
6. What is the thickness of material at hole (K) ?
7. What type of line is used to depict the temporary lugs?
8. What is the total tolerance on the 4.000 dimension of the end view?
9. What is the total tolerance on hole (G) ?
10. How many countersunk holes are there?
11. At what angle is surface (V) to the horizontal?
12. What is the distance between the .094 dia. holes?
13. Determine distances (2) through (26)
14. Locate the following lines or surfaces, placing the same letter on the appropriate views:
A. Top view; (L) (M) (N)
B. Front view; (P) (Q) (R)
C. End view; (S) (T) (U)

QUANTITY	500	
MATERIAL	CI	
SCALE	1/1	
DRAWN		DATE

DRIVE HOUSING | **A-56**

179

UNIT 35

WELDING DRAWINGS [1]

The primary importance of welding is to unite various pieces of metals so that they will properly operate as a unit structure to support the loads to be carried. In order to design and build such a structure, which will be both economical and efficient, a knowledge of welding and available steel shapes is essential.

The introduction of welding symbols on a drawing enables the designer to indicate clearly the type and size of weld required to meet his design requirements, and it is becoming increasingly important for the designer to indicate the required type of weld correctly. Points which must be made clear are the type of weld, the joint penetration, the weld size, the root gap (if any), and the degree of penetration required. These points can be clearly indicated on the drawing by the welding symbol.

WELDING SYMBOLS

Welding symbols are a shorthand language. They save time and money and serve

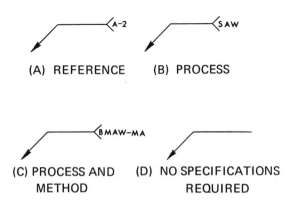

(A) REFERENCE   (B) PROCESS

(C) PROCESS AND   (D) NO SPECIFICATIONS
     METHOD   REQUIRED

Fig. 35-1 Location of Specifications, Processes, and Other References on Welding Symbols. (Courtesy Canadian Welding Bureau and American Welding Society)

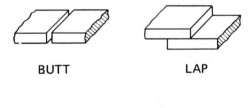

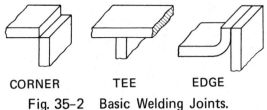

Fig. 35-2 Basic Welding Joints.

to insure understanding and accuracy. It is desirable that they should be a universal language; and for this reason the symbols of the American Welding Society have been adopted.

Any joint of welding of which is indicated by a symbol will always have an *arrow side* and an *other side.* Accordingly, the words *arrow side, other side,* and *both sides* are used to locate the weld with respect to the joint. The *tail* of the symbol is used for designating

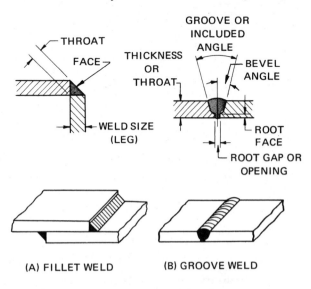

(A) FILLET WELD   (B) GROOVE WELD

Fig. 35-3 Basic Welding Nomenclature.
(Courtesy Canadian Welding Bureau)

181

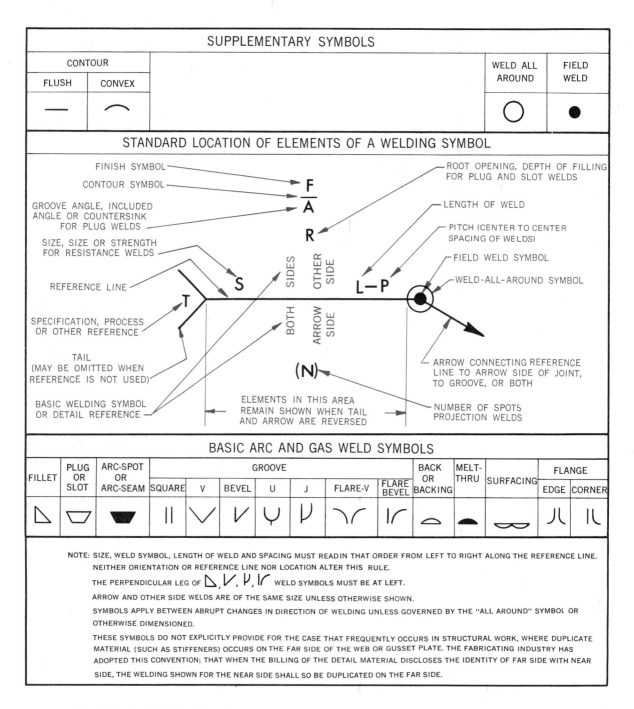

Fig. 35-4   Welding Symbols. (Courtesy Canadian Welding Bureau and American Welding Society)

the welding specifications, procedures, or other supplementary information to be used when making the weld. The notation to be placed on the *tail* of the symbol is to indicate the process, the type of filler metal to be used and whether or not peening or root chipping is required. If notations are not used, the *tail* of the symbol may be omitted.

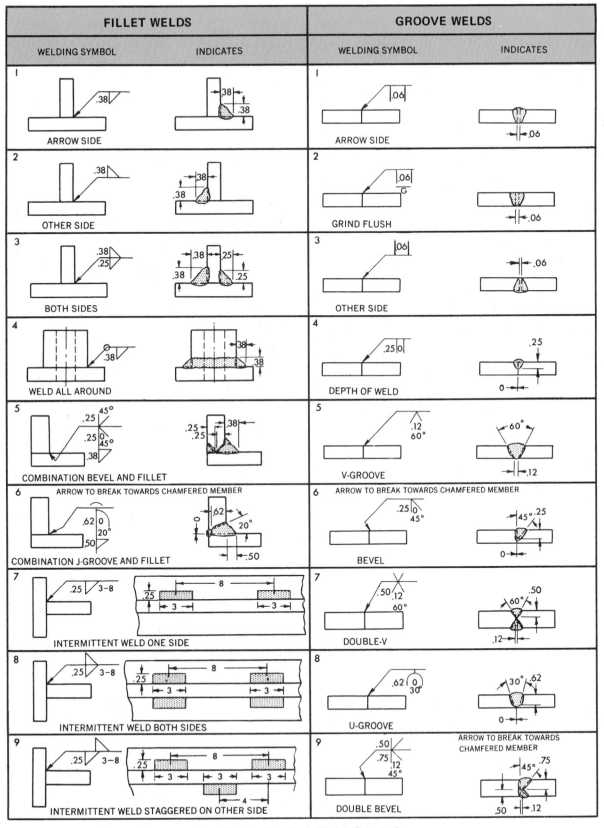

Fig. 35-5 Typical Weld Symbols.

Welds on the *arrow side* of the joint are shown by placing the weld symbol on the bottom side of the reference line. Welds on the *other side* of the joint are shown by placing the weld symbol on the top side of the reference line. Welds on *both sides* of the joint are shown by placing the weld symbol on both sides of the reference line. A weld extending completely around a joint is indicated by means of a *weld-all-around* symbol placed at the intersection of the *reference line* and the *arrow.*

Field welds (welds not made in the shop or at the initial place of construction) are indicated by means of the *field weld* symbol placed at the intersection of the *reference line* and the *arrow.*

Only the basic *fillet* and *groove welds* will be considered at this time.

Fillet Welds

1. Dimensions of fillet welds are shown on the same side of the reference line as the weld symbol.

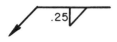

2. When both sides of a joint have the same size fillet welds, one or both may be dimensioned.

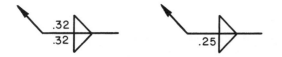

3. When both sides of a joint have different size fillet welds, both are dimensioned.

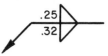

4. The dimension does not need to be shown when a general note is placed on the drawing to specify the dimension of fillet welds, and all the welds have dimensions governed by the note. Such a note might state, *ALL FILLET WELDS .32 UNLESS OTHERWISE NOTED.*

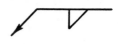

5. The *size* of a fillet weld is shown to the left of the weld symbol.

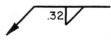

6. The *length* of a fillet weld, when indicated on the welding symbol, is shown to the right of the weld symbol.

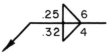

7. The *pitch* (center-to-center spacing) of an intermittent fillet weld is shown as the distance between centers of increments on one side of the joint. It is shown to the right of the length dimension.

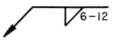

8. Staggered intermittent fillet welds are illustrated by staggering the weld symbols.

Groove Welds

1. Dimensions of groove welds are shown on the same side of the reference line as the weld symbol.

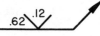

2. When both sides of a double-groove weld have the same dimensions, one or both may be dimensioned

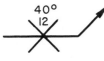

3. When both sides of a double-groove weld differ in dimensions, both are dimensioned.

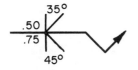

4. Groove welds do not need to be dimensioned when a general note appears on the drawing to govern the dimensions of groove welds, such as *ALL V-GROOVE WELDS ARE TO HAVE A 60° ANGLE UNLESS OTHERWISE NOTED.*

5. The size of groove welds is shown to the left of the weld symbol.

6. When the single-groove and symmetrical double-groove welds extend completely through the member or members being joined, the size of the weld need not be shown on the welding symbol.

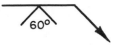

7. When the groove welds extend only partly through the member being joined, the size of the weld is shown on the weld symbol.

REFERENCES AND SOURCE MATERIAL

1. Canadian Welding Bureau and American Welding Society.

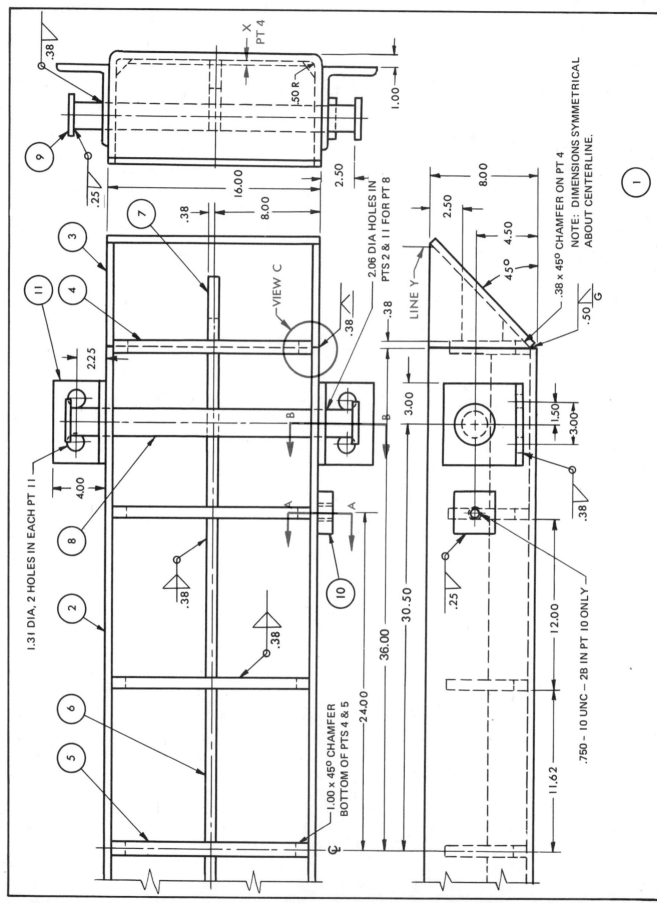

SKETCHING ASSIGNMENT

On the graph sections, complete the sketches and darken in welds. Square sizes are designated.

QUESTIONS

1. How many 1.31 dia. holes are there in the complete assembly?

2. How deep is the .750 tapped hole?

3. What class of fit is required for the .750 tapped hole?

4. What is the greatest overall width of the assembly?

5. Determine the distance from line **Y** to the centerline of the assembly.

6. Determine distance **X**.

7. What is the developed width of part 2? Use inside dimensions of channel.

8. What is the length of (A) pt. 2, (B) pt. 4, (C) pt. 7, (D) pt. 8?

9. What is the allowance for fitting on the length of pt. 5?

10. What is the difference between pt. 4 and pt. 5?

11. How many 1.00 chamfers are needed?

12. What does **G** mean on .50 weld symbol?

13. What type of weld is used to fasten pt. 11 to pt. 2?

14. What type of weld is used to fasten pt. 3 to pt. 2 at the sides?

15. How many parts are in the assembly?

.50 SQUARES

1.00 SQUARES

1.00 SQUARES

ENLARGED VIEW C

SECTION B-B

SECTION A-A

ANSWERS

1 _____
2 _____
3 _____
4 _____
5 _____
6 _____
7 _____
8 A _____
  B _____
  C _____
  D _____
9 _____
10 _____
11 _____
12 _____
13 _____
14 _____
15 _____

QTY	ITEM	MATL	DESCRIPTION	PT NO
4	LOCATING ANGLE	ST	L 6.00 x 4.00 x .50 x 6.00 LG	11
2	GROUND PAD	ST BAR	1.00 x 3.00 x 3.00	10
4	RETAINER	ST PL	.50 x 3.00 DIA	9
2	DRAW BAR	ST RD	2.00 DIA. x _____ LG	8
2	GUSSET	ST BAR	.75 x 3.00 x _____ LG	7
6	GUSSET	ST BAR	.75 x 3.00 x 11.25	6
5	GUSSET	ST BAR	.75 x 6.00 x 14.93	5
2	GUSSET	ST BAR	.75 x 6.00 x _____	4
2	END PLATE	ST PL	.50 x 10.62 x 25.50	3
1	BASE	ST PL	.50 x _____ W x _____ LG	2
1	SKID ASSY			1
QTY	ITEM	MATL	DESCRIPTION	PT NO

BASE SKID

A-57

187

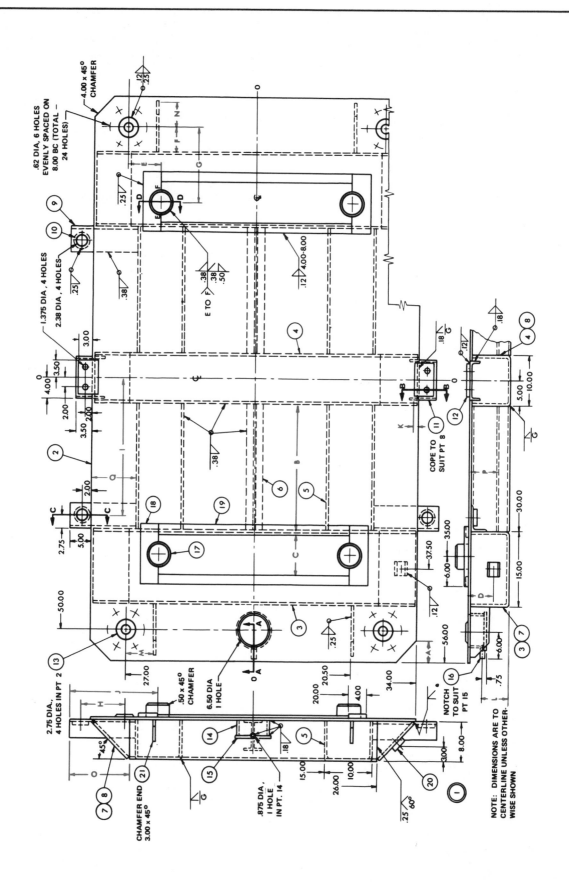

1/2- INCH SQUARES

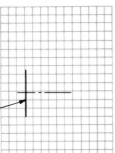

SECTION A-A

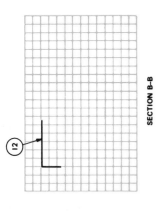

SECTION B-B

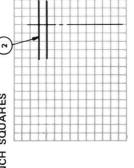

SECTION C-C

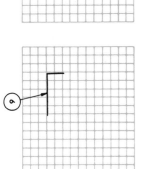

SECTION D-D

PT NO	QTY	ITEM	MATL	DESCRIPTION
21	4	RIB	ST	.50 x 4.00 x 10.00
20	1	GROUND PAD	ST	DWG A-4158
19	4	RETAINER	ST	.50 x 2.00 x _____ LG
18	4	RETAINER	ST	.75 x 4.00 x _____ LG
17	4	BUMPER PIN	ST	5.00 DIA x 3.00 LG
16	1	NIPPLE	ST	.50 IPS x 3.00 LG THD ONE END
15	1	END PLATE	ST	7.20 DIA x .50 THK
14	1	SUMP	ST	6.00 IPS x 3.00 LG
13	4	FLANGE	ST	WELD-IN FLANGE 3.00 IPS
12	2	SUPPORT	ST	.75 x 4.50 x 7.00
11	2	SUPPORT	ST	8.00 ⌴ 11.5 x 7.00 LG
10	4	SUPPORT	ST	3.50 DIA x 1.00 THK
9	4	DRAW BAR	ST	∟ 5.00 x 3.50 x .50 x _____ LG
8	2	RIB END	ST	.50 x 10.50 x 20.00
7	4	RIB END	ST	.50 x 10.50 x 25.00
6	2	RIB	ST	6.00 B 12.50 x 25.00 LG
5	4	RIB	ST PL	.50 x _____ x _____ LG
4	1	RIB	ST PL	.50 x _____ x _____ LG
3	2	RIB	ST PL	.50 x _____ x _____ LG
2	1	BASE PLATE	ST PL	.50 x _____ x _____ LG
1	1	BASE ASSEMBLY		
PT NO	QTY	ITEM	MATL	DESCRIPTION

DRAWN

BASE ASSEMBLY   **A-58**

SCALE 1/10

QUESTIONS

1. Determine dimensions **A** to **P.**

2. What is the overall (A) length; (B) width, and (C) height of the complete assembly?

3. Assuming the formed channels to have square inside corners, what is the width of the material used to make pt. 3? (Use inside travel.)

4. What is the (A) width; (B) length; of the baseplate, pt. 2?

5. If steel weighs .281 pounds per cu. in., what is the weight of the baseplate, pt. 2? (Disregard holes.)

6. Determine the distance from the bottom of the I-beam to the bottom of the base assembly.

7. What is the angle between the .62 dia. holes?

8. If 6-in. pipe has an O.D. of 6.62, what distance does pt. 15 project beyond pt. 14?

9. Determine the distance pt. 16 projects into pt. 14.

10. What is the area enclosed by parts 17, 18, and 19?

SKETCHING ASSIGNMENTS

1. In the areas provided, sketch sections A–A, B–B, C–C, and D–D, and darken in the welds.

2. Sketch one-half the development of pt. 8, showing dimensions, and indicate bend lines.

ANSWERS

1 A _____   O _____

  B _____   P _____

  C _____   2 A _____

  D _____   B _____

  E _____   C _____

  F _____   3 _____

  G _____   4 A _____

  H _____   B _____

  I _____   5 _____

  J _____   6 _____

  K _____   7 _____

  L _____   8 _____

  M _____   9 _____

  N _____   10 _____

189

CAMS

The cam plays an invaluable part in the design of automatic machinery because it is possible by using cams to impart any desired motion to another mechanism.

A cam is a rotating, oscillating, or reciprocating machine element which has a surface or groove formed to impart special or irregular motion to a second part called a follower. The follower rides against the curved surface of the cam. The distance that the follower rises and falls in a definite period of time is determined by the shape of the cam profile.

Types of Cams

The type and shape of cam is dictated by the required relationship of the parts and the

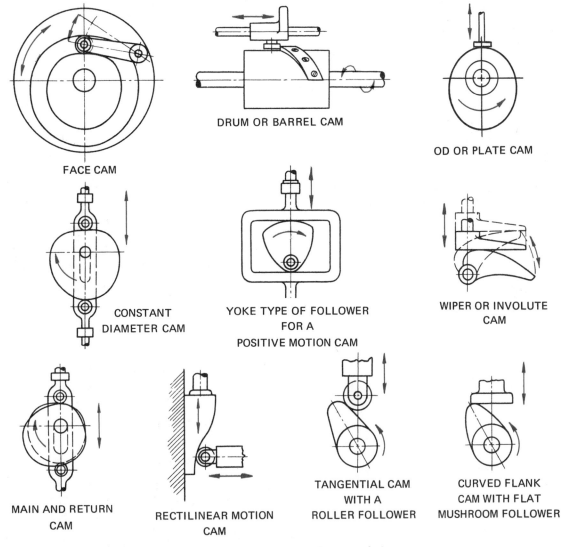

FACE CAM

DRUM OR BARREL CAM

OD OR PLATE CAM

CONSTANT DIAMETER CAM

YOKE TYPE OF FOLLOWER FOR A POSITIVE MOTION CAM

WIPER OR INVOLUTE CAM

MAIN AND RETURN CAM

RECTILINEAR MOTION CAM

TANGENTIAL CAM WITH A ROLLER FOLLOWER

CURVED FLANK CAM WITH FLAT MUSHROOM FOLLOWER

Fig. 36-1   Common Types of Cams.

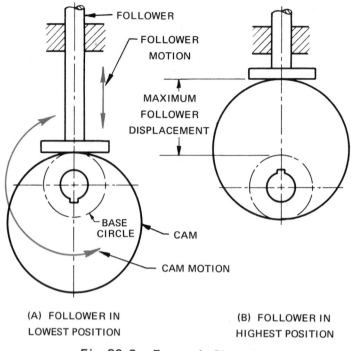

(A) FOLLOWER IN LOWEST POSITION

(B) FOLLOWER IN HIGHEST POSITION

Fig. 36-2 Eccentric Plate Cam.

motions of both. The type of cams which are generally used are either radial or cylindrical. The follower of a radial or face cam moves in a plane perpendicular to the axis of the cam, while in the cylindrical type of cam the movement of the follower is parallel to the cam axis.

A simple OD (outside diameter) or plate cam is shown in figure 36-2. The hole in the plate is bored off-center so that as it revolves

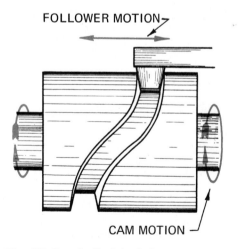

Fig. 36-3   Cylindrical Grooved Cam (Drum or Barrel Type).

it causes the follower to move up and down. The follower can be of any type that will roll or slide on the surface of the cam. The follower used with this cam is called a flat face follower.

The cam shown in figure 36-3 is a cylindrical, grooved cam that transmits motion transversely to a lever connected to a conical follower which rides in the groove as the cam revolves.

Cam Displacement Diagrams

In preparing cam drawings, a cam displacement diagram is drawn first to plot the motion of the follower. The curve on the drawing represents the path of the follower, not the face of the cam. The diagram may be any convenient length, but often it is drawn equal to the circumference of the base circle of the cam and the height is drawn equal to the follower displacement. The lines drawn on the motion diagram are shown as radial lines on the cam drawing, and sizes are transferred from the motion diagram to the cam drawing.

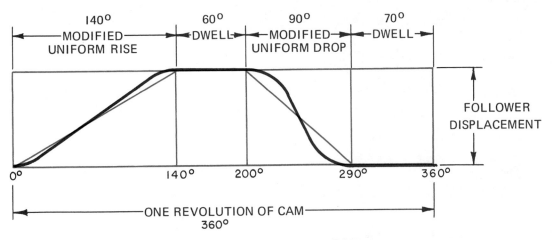

Fig. 36-4 Cam Displacement Diagram.

Figure 36-4 shows a cam displacement diagram having a modified uniform type of motion plus two dwell periods. Most cam displacement diagrams have cam displacement angles of 360 degrees. For drum or cylindrical grooved cams, the displacement diagram is often replaced by the developed surface of the cam.

The drawing of the cylindrical feeder cam (drawing A-59) is a cylindrical, grooved cam. In addition to the working views of the cam, a development of the contour of the grooves is also shown. This development aids the machinist in scribing and laying out the contour of the cam action lobes on the surface of the cam blank preparatory to machining grooves.

Regardless of the type of cam or follower, the purpose of all cams is to impart motion to other mechanisms in various directions in order to actuate machines to do specific jobs. Several of the simpler types of cams are shown in figure 36-1.

QUESTIONS

1. Through what thickness of metal is hole ② drilled?

2. Locate line ④ in the right side view.

3. Locate line ⑪ in the cam displacement diagram.

4. Locate line ⑥ in another view.

5. Locate line ⑨ in the right side view.

6. What is the maximum permissible diameter of hole ② ?

7. Locate line ⑧ in the front view.

8. Locate line ⑨ in the left side view.

9. What is the total follower displacement of the cam follower for (A) the finishing cut, (B) the roughing cut?

10. How many surfaces are to be finished?

11. What is the total number of through holes?

12. Locate lines ㉚ to ㊵ on other view.

13. Determine distances Ⓐ to Ⓜ

14. Insert material used, in title block.

NOTE: UNLESS OTHERWISE SHOWN:
TOLERANCE ON TWO DECIMAL DIMENSIONS ± .02
TOLERANCE ON THREE DECIMAL DIMENSIONS ± .005
TOLERANCE ON ANGLES ± 30'

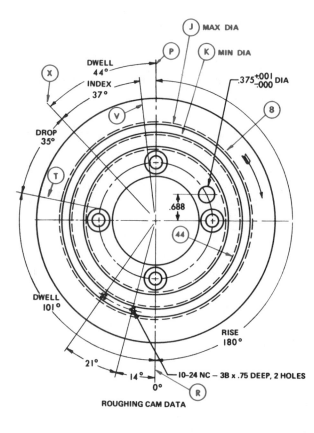

ANSWERS

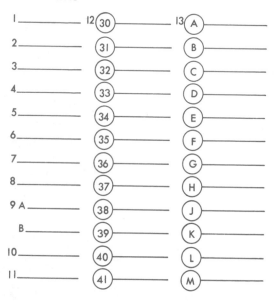

1 _____	12 ㉚ _____	13 Ⓐ _____
2 _____	㉛ _____	Ⓑ _____
3 _____	㉜ _____	Ⓒ _____
4 _____	㉝ _____	Ⓓ _____
5 _____	㉞ _____	Ⓔ _____
6 _____	㉟ _____	Ⓕ _____
7 _____	㊱ _____	Ⓖ _____
8 _____	㊲ _____	Ⓗ _____
9 A _____	㊳ _____	Ⓙ _____
B _____	㊴ _____	Ⓚ _____
10 _____	㊵ _____	Ⓛ _____
11 _____	㊶ _____	Ⓜ _____

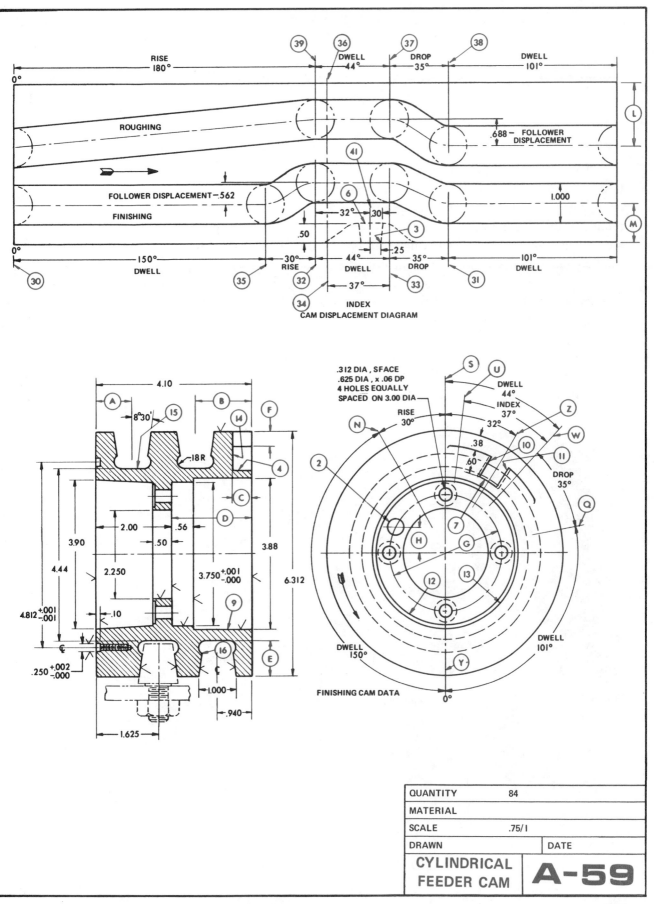

CAM DISPLACEMENT DIAGRAM

RISE 180°

ROUGHING

FOLLOWER DISPLACEMENT - .562

FINISHING

DWELL 44°

DROP 35°

DWELL 101°

.688 - FOLLOWER DISPLACEMENT

1.000

150° DWELL

30° RISE

44° DWELL

35° DROP

101° DWELL

.50

32°

30

.25

37°

INDEX

FINISHING CAM DATA

.312 DIA , SFACE
.625 DIA , x .06 DP
4 HOLES EQUALLY
SPACED ON 3.00 DIA

RISE 30°

DWELL 44°

INDEX 37°

32°

.38

.60

DROP 35°

DWELL 150°

DWELL 101°

4.10

8°30'

.18 R

2.00

.56

.50

3.90

4.44

2.250

3.750 +.001 -.000

4.812 +.001 -.001

.10

.250 +.002 -.000

1.000

.940

1.625

3.88

6.312

QUANTITY	84	
MATERIAL		
SCALE	.75/1	
DRAWN		DATE
CYLINDRICAL FEEDER CAM	A-59	

195

UNIT
37

GEARS

The function of a gear is to transmit motion, rotating or reciprocating, from one machine part to another and, where necessary, to reduce or increase the r.p.m. of a shaft. Gears are rolling cylinders or cones having teeth on their contact surfaces to insure position motion.

There are many kinds of gears, and they may be grouped according to the position of the shafts that they connect. Spur gears connect parallel shafts, bevel gears connect shafts whose axes intersect, and worm gears connect shafts whose axes do not intersect. A spur gear with a rack converts rotary motion to reciprocating or linear motion. The smaller of the two gears is known as the pinion.

Spur Gears

The standard spur gear tooth forms are developed from the basic involute curve system. Classically, the involute is described as the curve traced by a point on a taut string unwinding from a circle. This circle is called the base circle and is not a physical part of the gear and cannot be measured directly. The contact between mating involutes occurs along a line which is tangent to and crossing between the two base circles. This is the line of action.

Fig. 37-1   Gears.   (Courtesy Boston Gears)

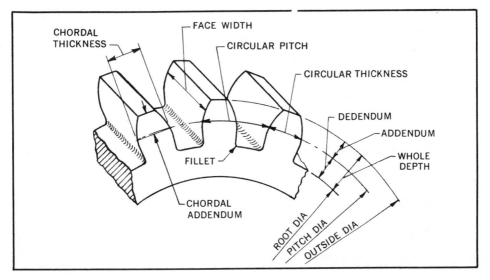

Fig. 37-2 Gear Tooth Terms.

The 14½ degree pressure angle has been used for many years and remains useful for duplicate or replacement gearing.

The 20-degree angle has become the standard for new gearing because of the smoother and quieter running characteristics and greater load-carrying ability.

The formulas for the 14½-degree and the 20-degree full depth teeth are identical.

Spur Gear Terms

Pitch Diameter. The diameter of an imaginary circle on which the gear tooth is designed.

Diametral Pitch. A ratio equal to the number of teeth on a gear for every inch of pitch diameter.

Addendum. The radial distance from the pitch circle to the top of the tooth.

Dedendum. The radial distance from the pitch circle to the bottom of the tooth.

Whole Depth. The overall height of the tooth.

PITCH	$14\frac{1}{2}°$	20°
4		
8		
12		
20		

Fig. 37-4 Actual size of 14½° and 20° Gear Teeth of Different Diametral Pitches.

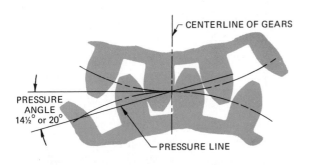

Fig. 37-3 Meshing of Gear Teeth.

TERM	SYMBOL	FORMULA
PITCH DIAMETER	D	$D = \dfrac{N}{P}$
NUMBER OF TEETH	N	$N = D \times P$
DIAMETRAL PITCH	P	$P = \dfrac{N}{D}$
ADDENDUM	A	$A = \dfrac{1}{P}$
DEDENDUM.	B	$B = \dfrac{1.157}{P}$
WHOLE DEPTH	WD	$WD = \dfrac{2.157}{P}$
CLEARANCE	C	$C = \dfrac{.157}{P}$
OUTSIDE DIAMETER	OD	$OD = D \times 2A = \dfrac{N + 2}{P}$
ROOT DIAMETER	RD	$RD = D - 2B = \dfrac{N - 2.314}{P}$
CIRCULAR PITCH	CP	$CP = \dfrac{3.1416D}{N} = \dfrac{3.1416}{P}$
CIRCULAR THICKNESS	T	$T = \dfrac{3.1416D}{2N} = \dfrac{1.57}{P}$
CHORDAL THICKNESS	Tc	$Tc = D\ SIN\left(\dfrac{90^{\circ}}{N}\right)$
CHORDAL ADDENDUM	Ac	$Ac = A + \dfrac{T^2}{4D}$

Fig. 37-5   Spur Gear Symbols and Formulas.

Clearance.   The radial distance between the bottom of one tooth and the top of the mating tooth.

Outside Diameter.   The overall gear diameter.

Root Diameter.   The diameter at the bottom of the tooth.

Circular Pitch.   The distance measured from the point of one tooth to the corresponding point on the adjacent tooth on the circumference of the pitch diameter.

Circular Thickness.   The thickness of a tooth or space measured on the circumference of the pitch diameter.

Chordal Thickness.   The thickness of a tooth or space measured along a chord on the circumference of the pitch diameter.

Chordal Addendum.   Chordal addendum, also known as corrected addendum, is the perpendicular distance from the chord to outside circumference of the gear.

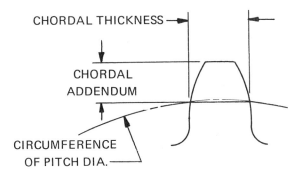

CHORDAL THICKNESS →

CHORDAL
ADDENDUM

CIRCUMFERENCE
OF PITCH DIA. ⏌

Fig. 37-6   Chordal Addendum and Thickness.

Chordal Thickness and Corrected Addendum

After the gear teeth have been milled or generated, the width of the tooth space and the thickness of the tooth, measured on the pitch circle, should be equal.

Instead of measuring the curved length of line known as *circular thickness of tooth,* it is more convenient to measure the length of the straight line *(chordal thickness)* which connects the ends of that arc.

The *corrected* or *chordal addendum* is the radial distance extending from the addendum circle to the chord.

A gear tooth vernier caliper may be used to measure accurately the thickness of a gear tooth at the pitch line. To use the gear tooth vernier which measures only a straight line or chordal distance, it is necessary to set the tongue to the computed *chordal addendum,* and then measure the chordal thickness.

Working Drawings of Spur Gears

The working drawings of gears which are normally cut from blanks are not complicated. A sectional view is sufficient unless a front view is required to show web or arm details. Since the teeth are cut to shape by cutters, they need not be shown in the front view.

The CSA recommends that a solid line be used for the outside and root circles, and a centerline for the pitch circle. The ASA recommends the use of phantom lines for the outside and root circles. In the section view the root and outside circles are shown as solid lines.

The dimensioning for the gear is divided into two groups because the finishing of the gear blank and the cutting of the teeth are separate operations in the shop. The gear blank dimensions are shown on the drawing while the gear tooth information is given in a table.

QUESTIONS

1. What is the hub diameter?

2. What is the maximum thickness of the spokes?

3. What is the average width of the spokes?

4. How many surfaces indicate that allowance must be added to pattern for finishing?

5. Determine distance (J) for the pattern. Assume .10 is allowed on pattern for each surface to be finished.

6. Determine distance (K) for the pattern.

7. What is the outside diameter of the pattern?

8. Determine distance (L) for the pattern.

9. What is the width of the pattern?

10. What is the diametral pitch of the gear?

11. Calculate the center-to-center distance if this gear were to mesh with a pinion having (A) 24 teeth, (B) 36 teeth, (C) 32 teeth.

12. Calculate distances (E) through (H) .

13. Calculate the following: addendum, dedendum, circular pitch, and root diameter.

14. Complete the missing information in the cutting data table.

ANSWERS

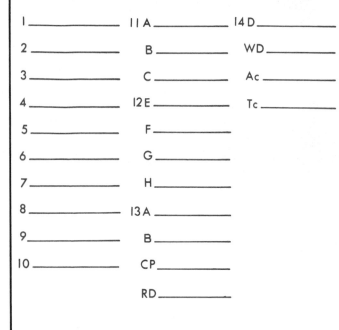

1 _____	11 A _____	14 D _____
2 _____	B _____	WD _____
3 _____	C _____	Ac _____
4 _____	12 E _____	Tc _____
5 _____	F _____	
6 _____	G _____	
7 _____	H _____	
8 _____	13 A _____	
9 _____	B _____	
10 _____	CP _____	
	RD _____	

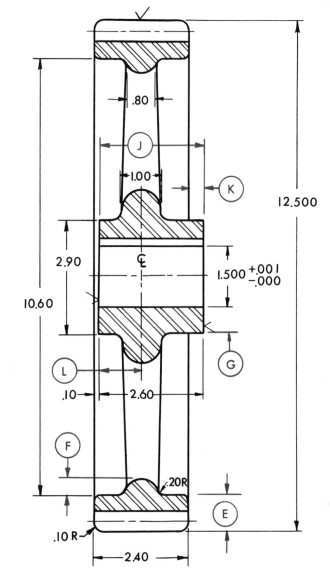

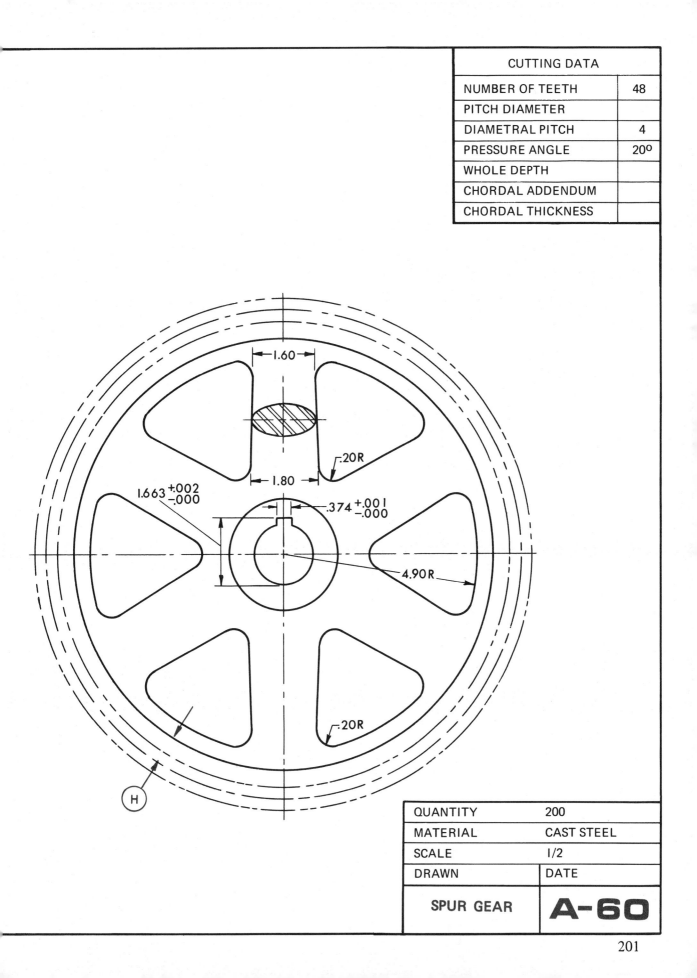

CUTTING DATA

NUMBER OF TEETH	48
PITCH DIAMETER	
DIAMETRAL PITCH	4
PRESSURE ANGLE	20°
WHOLE DEPTH	
CHORDAL ADDENDUM	
CHORDAL THICKNESS	

1.60

.20R

1.80

$1.663^{+.002}_{-.000}$

$.374^{+.001}_{-.000}$

4.90R

.20R

H

QUANTITY	200
MATERIAL	CAST STEEL
SCALE	1/2
DRAWN	DATE
SPUR GEAR	**A-60**

UNIT 38

BEVEL GEARS

Drawings of bevel gears may be more easily interpreted and understood as a result of having a working knowledge of the parts, principles, and formulas underlying spur gears.

Spur gears transmit motion by or through shafts that are parallel and in the same plane, while bevel gears transmit motion between shafts which are in the same plane but whose axes would meet if extended. Theoretically, the teeth of a spur gear may be said to be built about the original frictional cylindrical surface known as *pitch circle,* while the teeth of a bevel gear are formed about the frustum of the original conical surface called *pitch cone.*

One type of bevel gear which is commonly used is the miter gear. The term miter gear refers to a pair of bevel gears of the same size that transmit motion at right angles.

While any two spur gears of the same diametral pitch will mesh, this is not true of bevel gears except for miter gears. On each pair of mating bevel gears, the diameters of the gears determine the angles at which the teeth are cut.

Working drawings of bevel gears, like spur gears, give only the dimensions of the bevel gear blank. Cutting data for the teeth is given in a note or table. A single section view is used unless a second view is required to show such details as spokes. Sometimes both the bevel gear and pinion are drawn together. Dimensions and cutting data depend on the method used in cutting the teeth, but the information in figure 38-1 is commonly used.

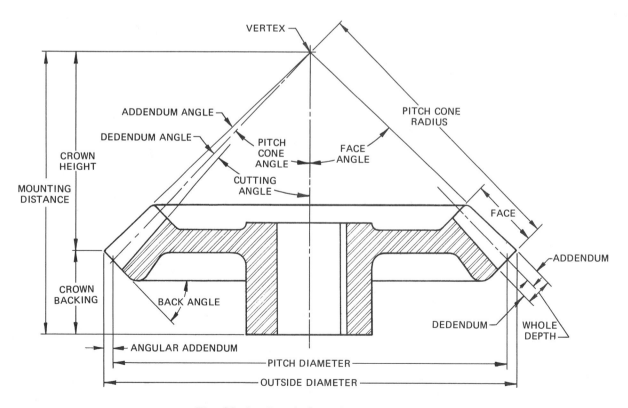

Fig. 38-1   Bevel Gear Nomenclature.

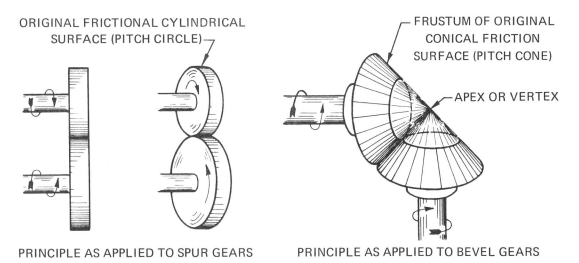

PRINCIPLE AS APPLIED TO SPUR GEARS   PRINCIPLE AS APPLIED TO BEVEL GEARS

Fig. 38-2   Comparing Pitch Circles of Spur Gears With Pitch Cones of Bevel Gears.

TERM	FORMULA
Addendum, dedendum, whole depth, pitch diameter, diametral pitch, number of teeth, circular pitch, chordal thickness, circular thickness	Same as for spur gears. Refer to Figure 37-5.
Pitch cone angle (Pitch angle)	$\text{Tan pitch angle} = \dfrac{\text{D of gear}}{\text{D of pinion}} = \dfrac{\text{N of gear}}{\text{N of pinion}}$
Addendum angle	$\text{Tan addendum angle} = \dfrac{\text{Addendum}}{\text{Pitch cone radius}}$
Dedendum angle	$\text{Tan dedendum angle} = \dfrac{\text{Dedendum}}{\text{Pitch cone radius}}$
Face angle	Pitch cone angle plus addendum angle
Cutting angle	Pitch cone angle minus dedendum angle
Back angle	Same as pitch cone angle
Angular addendum	Cosine of pitch cone angle x addendum
Outside diameter	Pitch diameter plus two angular addendum
Crown height	Divide 1/2 the outside diameter by the tangent of the face angle
Face width	1½ to 2½ times the circular pitch
Chordal addendum	$A + \dfrac{T^2 \times \text{Cos pitch cone angle}}{4D}$

Fig. 38-3   Bevel Gear Formulas.

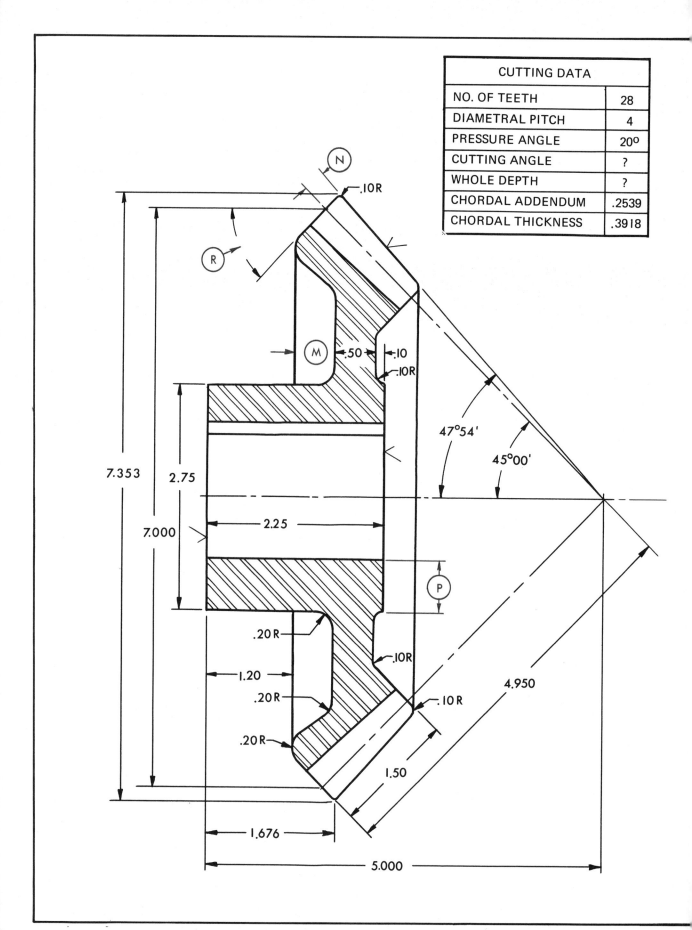

CUTTING DATA

NO. OF TEETH	28
DIAMETRAL PITCH	4
PRESSURE ANGLE	20°
CUTTING ANGLE	?
WHOLE DEPTH	?
CHORDAL ADDENDUM	.2539
CHORDAL THICKNESS	.3918

N

R

.10R

M .50 .10

.10R

47°54'

45°00'

7.353

2.75

7.000

2.25

P

.20R

1.20

.20R

.10R

.10 R

.20R

4.950

1.50

1.676

5.000

QUESTIONS

1. The draftsman neglected to put on angle dimension (R). What should it be?

2. How many finished surfaces are indicated?

3. List those dimensions shown on the section drawing which are not used by the patternmaker. Assume that the hole will not be cored.

4. What is the pitch cone angle?

5. What is the pitch diameter?

6. Indicate those dimensions on the section drawing which the machinist would use to machine the blank before the teeth are cut.

7. What is the depth of teeth at the large end?

8. What is the pitch cone radius?

9. What is the face angle?

10. What is the addendum angle?

11. What is the cutting angle?

12. What is the mounting distance?

13. What is the crown height?

14. What is the angular addendum?

15. What is the face width?

16. Determine dimensions (M),(N),(P).

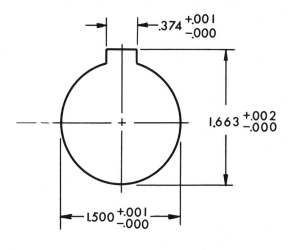

.374 $^{+.001}_{-.000}$

1.663 $^{+.002}_{-.000}$

1.500 $^{+.001}_{-.000}$

ANSWERS

1 _____ 5 _____ 10 _____

2 _____ 6 _____ 11 _____

3 _____ _____ 12 _____

_____ _____ 13 _____

_____ _____ 14 _____

_____ _____ 15 _____

_____ 7 _____ 16M _____

_____ 8 _____ N _____

4 _____ 9 _____ P _____

QUANTITY	50	
MATERIAL	CAST STEEL	
SCALE	1/1	
DRAWN		DATE
MITER GEAR		A-61

GEAR TRAINS

Center Distance. The center distance between the two shaft centers is determined by adding the pitch diameter of the two gears together and dividing the sum by 2.

Example: An 8 diametral pitch, 24-tooth pinion mates with an 8 diametral pitch, 96-tooth gear. Find the center distance.

$$\text{Pitch dia. of pinion} = \frac{24 \text{ teeth}}{8P} = 3.000 \text{ in.}$$

$$\text{Pitch dia. of gear} = \frac{96 \text{ teeth}}{8P} = 12.000 \text{ in.}$$

Sum of the two pitch dia. = 3.000 in. + 12.000 in. = 15.000 in.

Center distance = ½ sum of the two pitch dia.

$$= \frac{15.000 \text{ in.}}{2} = 7.500 \text{ in.}$$

Ratio. The ratio of gears is a relationship between any of the following:

- The r.p.m. of the gears
- The number of teeth on the gears
- The pitch dia. of the gears

The ratio is obtained by dividing the larger value of any of the three by the corresponding smaller value.

Examples:

1. A gear rotates at 90 r.p.m. and the pinion at 360 r.p.m.

$$\text{Ratio} = \frac{360}{90} = 4 \text{ or ratio} = 4:1$$

2. A gear has 72 teeth; the pinion, 18 teeth.

$$\text{Ratio} = \frac{72}{18} = 4 \text{ or ratio} = 4:1$$

3. A gear with a pitch dia. of 8.500 in. meshes with a pinion having a pitch dia. of 2.125 in.

$$\text{Ratio} = \frac{\text{D of gear}}{\text{D of pinion}} = \frac{8.500}{2.125} = 4 \text{ or}$$

$$\text{ratio} = 4:1$$

Determining Pitch Dia. and Outside Dia. (OD). The pitch diameter of a gear can readily be found if the number of teeth and diametral pitch are known. The OD is equal to the pitch diameter plus two addendums. The addendum for a 14½- or 20-degree spur gear tooth is equal to 1/P.

Examples:

1. A 14½-degree spur gear has a P of 4 and 34 teeth.

$$\text{Pitch dia.} = \frac{N}{P} = \frac{34}{4} = 8.500 \text{ in.}$$

$$OD = D + 2A = 8.500 + 2(¼) = 9.000 \text{ in.}$$

2. The OD of a 14½-degree spur gear is 6.500 in. The gear has 24 teeth.

$$OD = \frac{N + 2}{P} = \frac{24 + 2}{P} = \frac{26}{P} = 6.500 \text{ in.}$$

$$P = \frac{26}{6.500} = 4$$

$$\text{Addendum} = \frac{1}{P} = \frac{1}{4} = .250 \text{ in.}$$

$$\text{Pitch Dia.} = OD - 2A = 6.500 - 2(.250)$$
$$= 6.000 \text{ in.}$$

Problems in Spur Gear Calculations

1. Given N = 24, P = 6, 20-degree involute spur gear, find: D, A, WD, OD, CP.

2. Given D = 5.000 in., N = 40, 14½-degree involute spur gear, find: A, P, RD, OD, CP.

3. Given D = 6.000 in., A = .250, 20-degree involute spur gear, find: P, N, WD, OD, CP.

4. Given center-to-center distance between gear and pinion of 6.000 in., 36 teeth on the gear, 12 teeth on the pinion, find for gear and pinion: D, P, A, OD, CP, RD.

Motor Drive

Drawing A-63 shows a motor drive similar to the type used to operate a load-ratio control switch on a power transformer.

The load-ratio control switch is operated by a 1/6-hp. motor with a speed of 1,080 r.p.m. The shaft speed at the switch is reduced to 9 r.p.m. by a series of spur and miter gears.

When the operator pushes a button, the motor is activated until such time as the circuit breaker pointer rotates 90 degrees and one of the arms depresses the roller and breaks the circuit. During this time the load-ratio control switch shaft will rotate 360 degrees, moving the contactor in the load-ratio control switch one position. This will be shown on the dial by the position indicator.

To simplify the assembly, only the pitch diameters of the gears are shown and much of the hardware has been omitted.

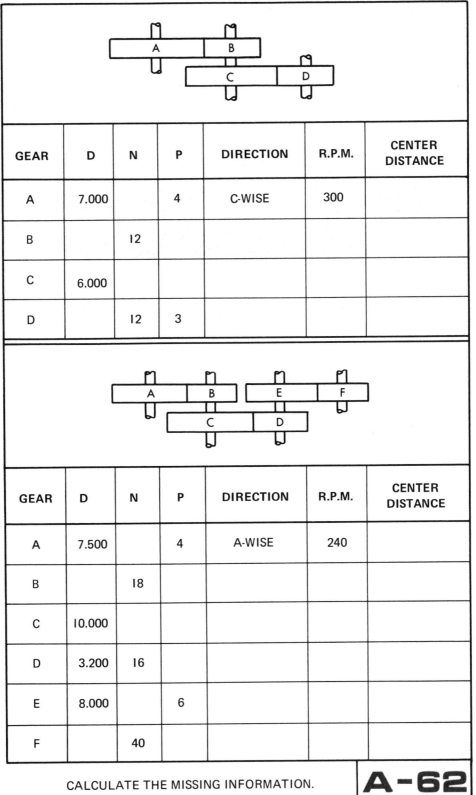

GEAR	D	N	P	DIRECTION	R.P.M.	CENTER DISTANCE
A	7.000		4	C-WISE	300	
B		12				
C	6.000					
D		12	3			

GEAR	D	N	P	DIRECTION	R.P.M.	CENTER DISTANCE
A	7.500		4	A-WISE	240	
B		18				
C	10.000					
D	3.200	16				
E	8.000		6			
F		40				

CALCULATE THE MISSING INFORMATION.

A-62

GEAR DATA				
GEAR	NO. OF TEETH	PITCH DIA.	P	R.P.M.
G_1	24		20	
G_2		4.80		
G_3	20	1.00		
G_4	100			
G_5		1.00	20	
G_6		6.00		
G_7	18			
G_8		7.20		
G_9	72		20	
G_{10}		2.50	10	
G_{11}	25			

SHAFT DATA			
SHAFT	GEARS ON SHAFT	R.P.M.	SHAFT * ROTATION
S_1			
S_2			
S_3			
S_4			
S_5			
S_6			
S_7			A WISE

* AS VIEWED FROM FRONT OR BOTTOM OF MOTOR DRIVE ASSEMBLY

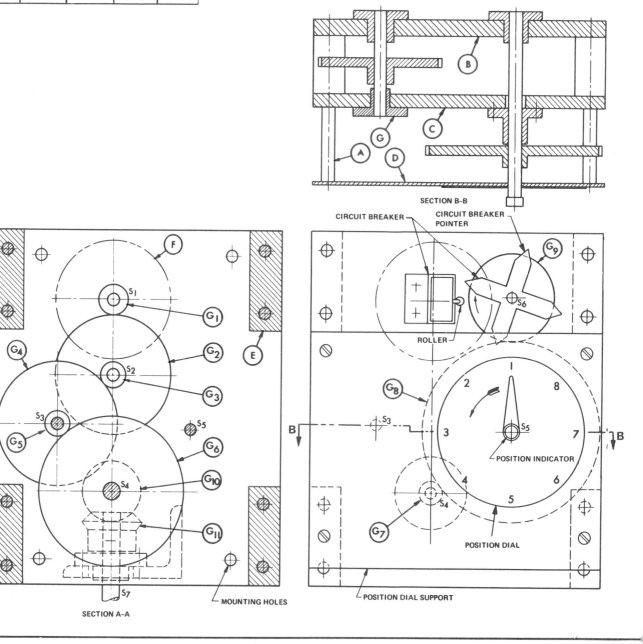

SECTION B-B

CIRCUIT BREAKER

CIRCUIT BREAKER POINTER

ROLLER

SECTION A-A

MOUNTING HOLES

POSITION INDICATOR

POSITION DIAL

POSITION DIAL SUPPORT

ASSIGNMENT

Complete the information shown in the gear and shaft tables

QUESTIONS

1. What are the names of parts (A) to (G) ?

2. How many spur gears are shown?

3. How many miter gears are shown?

4. How many gear shafts are there?

5. What is the ratio between the following gears? (A) G_1 and G_2, (B) G_3 and G_4, (C) G_5 and G_6, (D) G_7 and G_8, (E) G_8 and G_9, (F) G_{10} and G_{11}.

6. What is the center-to-center distance between the following shafts? (A) S_1 and S_2, (B) S_2 and S_3, (C) S_3 and S_4, (D) S_4 and S_5, (E) S_6 and S_6.

7. How many seconds does it take to turn the load ratio control switch one position?

8. How many seconds does it take the position indicator to move continuously from position 4 to position 7?

9. What is the r.p.m. ratio between the motor and the switch?

10. If the switch shaft S_7 rotates 1800 degrees, how many degrees does the position indicator rotate?

NOTE: PLEASE REFER TO PAGE **208** FOR INSTRUCTIONS REGARDING OPERATION.

ANSWERS	
IA	_____
B	_____
C	_____
D	_____
E	_____
F	_____
G	_____
2	_____
3	_____
4	_____
5A	_____
B	_____
C	_____
D	_____
E	_____
F	_____
6A	_____
B	_____
C	_____
D	_____
E	_____
7	_____
8	_____
9	_____
10	_____

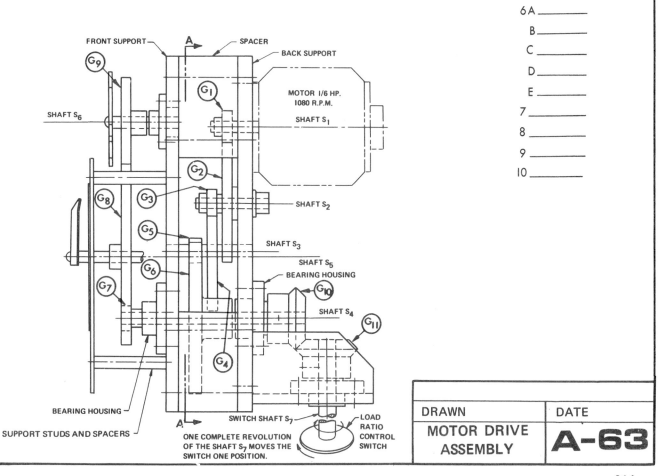

DRAWN	DATE
MOTOR DRIVE ASSEMBLY	**A-63**

TRUE POSITION DIMENSIONING [1]

Where features are located by means of rectangular coordinates, each dimension having its own tolerances, square- or rectangular-shaped tolerance zones result, figure 40-1A. The resultant tolerance zone is the .010 square. Since the actual position of the feature may be anywhere within the square, the maximum allowable variation from the desired position occurs at 45 degrees from the horizontal direction in which the tolerances are indicated. Therefore, the allowable variation on the diagonal is about 1.4 times the tolerance that would be used in coordinate tolerancing of rectangular coordinate dimensions without any increase in the maximum allowable variation. Such increase in tolerance often results in some reduction in cost of manufacture.

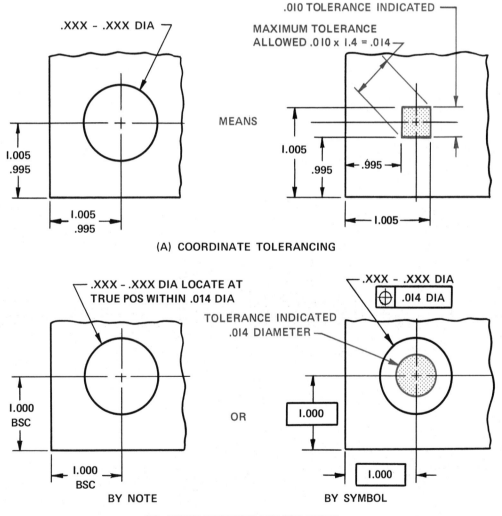

(A) COORDINATE TOLERANCING

(B) TRUE POSITION TOLERANCING

Fig. 40-1   A Comparison Between Coordinate and True Position Tolerancing.

True position dimensioning is the method which establishes the theoretical exact location of a feature, such as a hole or a boss. In this method, the locating dimensions represent the theoretical true position and are indicated by the abbreviation *BSC* (ASA) or *TP* (CSA) located either to the right or below the dimension. Tolerances do not apply directly to such dimensions, but permissible deviations

from the true position are indicated by a note attached to the feature size dimension, as shown in figure 40-1B. Either notes or symbols may be used on a drawing to indicate true position tolerancing.

True position tolerance is the diameter or width of a zone. It indicates the permissible deviation from true position point or plane, figure 40-2, since the most critical assembly condition exists when mating parts are made to the maximum material size, e.g. minimum hole diameter and maximum material condition (MMC). All true position tolerances apply on the MMC basis unless otherwise specified on the drawing.

The condition where tolerance of position or form must be met irrespective of where the feature lies within its tolerance is indicated on the drawing by the letters RFS (Regardless of Feature Size).

Analysis of a Two-Hole Pattern With True Position Tolerancing

A two-hole pattern lends itself to a simple analysis of true position tolerancing with MMC applied. Figure 40-3A shows the drawing requirements for the location and size of two holes. If the holes are exactly .500 in diameter (the maximum material condition, or the smallest size hole permitted by the drawing specification) and are centered exactly 2.000 apart, they will theoretically receive a gauge consisting of two round pins fixed in a plate if the pins are centered 2.000 apart and are .500 in diameter.

However, the limits specified on the drawing require the center distance between the holes to be from 1.993 to 2.007. If the hole is at minimum limit of size, the pins in the gauge would, therefore, have to be .007 smaller or .493 diameter, to enter the holes in their extreme positions, as shown in figure 40-3 C.

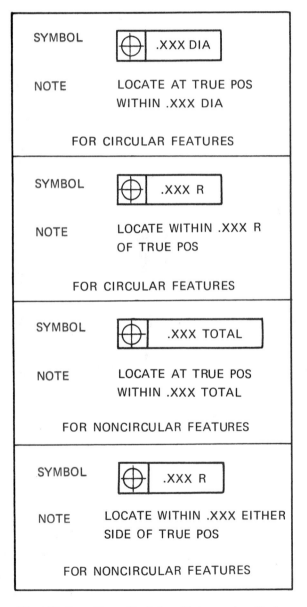

Fig. 40-2 True Position Tolerancing Notes and Symbols.

214

If the holes are exactly .500 in diameter but located at the maximum or minimum permissible center distance, the .493 diameter gauge pins would contact the sides of the holes. Neglecting gaugemakers' tolerances, the gauge pins would have to be .493 in diameter and the centers located exactly 2.000 apart. Thus, the holes in the part, if they are .500 diameter, will fit the gauge pins if located within the limit location specified on the drawing. If the holes are at maximum size, .505 diameter permitted by the drawing specification, they will accept the gauge.

GEOMETRICAL TOLERANCING [1]

A geometrical tolerance is the maximum permissible variation of form or position from that specified on the drawing, and represents the width or diameter of the tolerance zone, within which the edge, surface, or axis of the feature lies.

These tolerances control:

- The position or location of a feature

- The displacement of a feature about a common centerline, such as symmetry and concentricity

- The accuracy of a line or surface, such as straightness, flatness, roundness, contour, or profile

- The relationship of one line or surface with another, such as parallelism, squareness or perpendicularity, and angularity

Positional Tolerancing

A positional tolerance is the permissible variation in the specified position of a feature (surface, line or point) in relation to some other feature or datum.

Form Tolerancing

Form tolerances state how far the actual features of a part are permitted to vary from

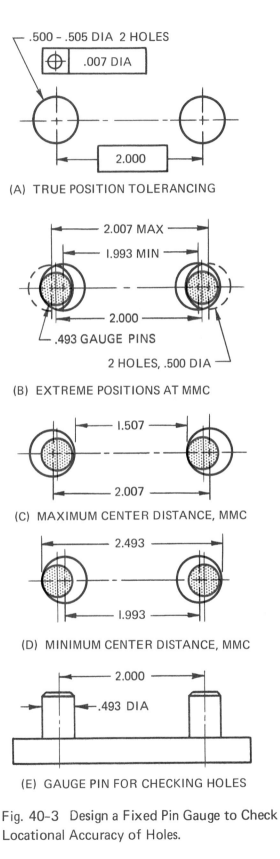

(A) TRUE POSITION TOLERANCING

.500 – .505 DIA 2 HOLES
⊕ .007 DIA
2.000

(B) EXTREME POSITIONS AT MMC

2.007 MAX
1.993 MIN
2.000
.493 GAUGE PINS
2 HOLES, .500 DIA

(C) MAXIMUM CENTER DISTANCE, MMC

1.507
2.007

(D) MINIMUM CENTER DISTANCE, MMC

2.493
1.993

(E) GAUGE PIN FOR CHECKING HOLES

2.000
.493 DIA

Fig. 40-3 Design a Fixed Pin Gauge to Check Locational Accuracy of Holes.

the perfect geometry specified or implied by drawings. The actual features of a part take various forms and shapes, and are defined by the use of geometric terms.

GEOMETRIC CHARACTERISTIC SYMBOLS			
CHARACTERISTIC			**SYMBOL**
FORM TOLERANCES	For Single Feature	Flatness	▱
		Straightness	—
		Roundness (Circularity)	○
		Cylindricity	⌭
		Profile of any line [1]	⌒
		Profile of any surface [1]	⌓
	For Related Features	Parallelism — ASA [2]	‖
		Parallelism — CSA	∕∕
		Perpendicularity (squareness)	⊥
		Angularity	∠
		Runout [3]	↗
Positional Tolerances		True Position	⊕
		Concentricity [4]	◎
		Symmetry [5]	≡
Supplementary Symbols		Maximum Material Size	Ⓜ
		Regardless of Feature Size	Ⓢ
		True Position	▭
		Datum A	−A−

1. Although included under "Form Tolerances," profile tolerances control size as well as form.

2. Parallel lines may be shown oblique.

3. Although included under "Form Tolerances," a runout tolerance controls position as well as form.

4. Where concentricity RFS applies, it is preferred that the runout symbol be used. Where concentricity at MMC applies, it is preferred that the true position symbol be used. Optionally, the inner circle may be filled solid.

5. Where symmetry applies, it is preferred that the true position symbol be used.

Fig. 40–4 Symbols for Position and Tolerance of Form.

Straightness. Straightness may be applied to edges, axes, or cylindrical surfaces.

Roundness and Cylindricity. When roundness is critical in assembly and function, a roundness tolerance should be specified. The roundness tolerance represents the maximum variation in radius from the virtual center. Cylindricity refers to a combination of roundness, straightness, and parallelism of a cylindrical surface.

Runout. Runout refers to the combined effect of eccentricity or out-of-roundness, and wobble. Wobble refers to the effect of an inclination of the axis of a cylindrical part. Runout tolerance may be referenced to the periphery of a cylindrical part. When runout consists mostly of eccentricity, or out-of-roundness, in relation to the flat face of a cylindrical part or disc, it is composed chiefly of wobble. This condition will exist in relation to an angular surface or a combination of both an angular surface and a flat surface.

Parallelism. When lines, surfaces, or axes are drawn parallel to one another, it is intended that they shall be parallel within normal process variations, but if control within close limits is necessary, a parallelism tolerance should be added. Usually, one surface or line is designated as the datum, and the controlled feature is related to it.

Squareness or Perpendicularity. Where the intended angle is 90 degrees, it is not necessary to specify the angle.

Application to Drawings [2]

Geometrical tolerances may be specified on drawings in note form or as tolerance symbols as shown in figure 40-5. When reference is made to a datum, the reference may be placed either before or after the tolerance, and

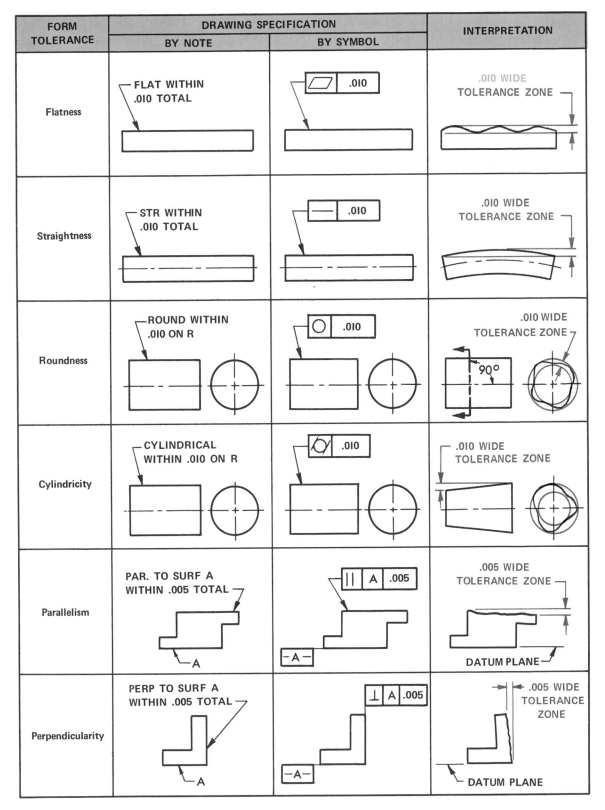

Fig. 40-5   Form Tolerances.

the datum may be more precisely identified by a descriptive word, as in the following examples:

LOCATE WITHIN .010 DIA DATUM HOLE A

SYMMETRICAL WITH WIDTH B WITHIN .005

PARALLEL WITH FACE C WITHIN .001

When required, the abbreviation MMC may be added to the tolerance, or to the datum, or both; for example:

STRAIGHT WITHIN .0015 MMC WITH DATUM A

CONCENTRIC WITHIN ZERO DIA MMC WITHIN DATUM B MMC

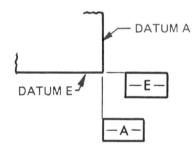

(A) DATUM-IDENTIFYING SYMBOL

(B) BASIC OR TRUE DIMENSION SYMBOL

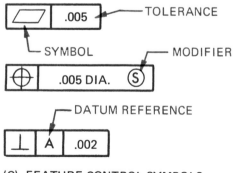

(C) FEATURE CONTROL SYMBOLS

Fig. 40-6   Positioning of Symbols.

Geometric Tolerance Symbols

These symbols are enclosed in a rectangular box, approximately .30 inch high, and as wide as necessary, with the applicable tolerance at datum, and a leader pointing to the feature to which it applies, figure 40-6.

When the modifier MMC is required, the symbol Ⓜ is used with the tolerance or datum to which it applies.

Datum-Identifying Symbol. The datum-identifying symbol consists of a rectangular frame containing the datum reference letter; the letter is preceded and followed by a dash. Each datum requiring identification on a drawing is assigned a different reference letter. The datum-identifying symbol is placed below the other symbols.

True Position Symbol. The symbolic means for labeling a basic or true position dimension is by enclosing each such dimension in a rectangular frame or box.

MMC and RFS Symbols. The symbols Ⓜ and Ⓢ are used to designate *Maximum Material Condition and Regardless of Feature Size*, respectively. The abbreviations MMC and RFS are used in notes.

Feature Control Symbol. A positional or form tolerance is stated by means of a feature control symbol which consists of a frame containing the geometric characteristic symbol followed by the permissible tolerance and, in some cases, by the modifier Ⓜ or Ⓢ. A vertical line separates the symbol from the tolerance.

Feature Control Symbol Incorporating Datum References. Where a positional or form tolerance must be related to a datum, this relationship is stated in the feature control symbol by placing the datum reference letter between the geometric symbol and the tolerance. Vertical lines separate these entries.

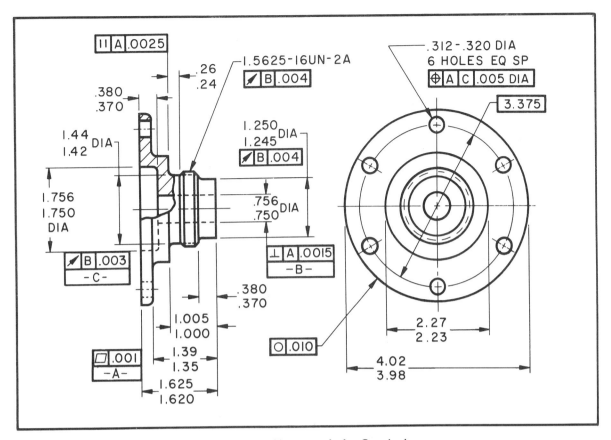

Fig. 40-7 Application of Geometric Characteristic Symbols. (Extracted from <u>USA Standard</u> <u>Drafting Practices</u>, Dimensioning and Tolerancing for Engineering Drawings (USASI Y14.5 — 1966), with the permission of the publisher, The American Society of Mechanical Engineers, United Engineering Center, 345 East 47th Street, New York, New York 10017.)

REFERENCES AND SOURCE MATERIAL

1. Extracted from *USA Standard Drafting Practices*, Dimensioning and Tolerancing for Engineering Drawings (USASI Y14.5 — 1966), with the permission of the publisher, The American Society of Mechanical Engineers, United Engineering Center), 345 East 47th Street, New York, N.Y. 10017.

2. Canadian Standards Association, *Bulletin* 78.2 (1967).

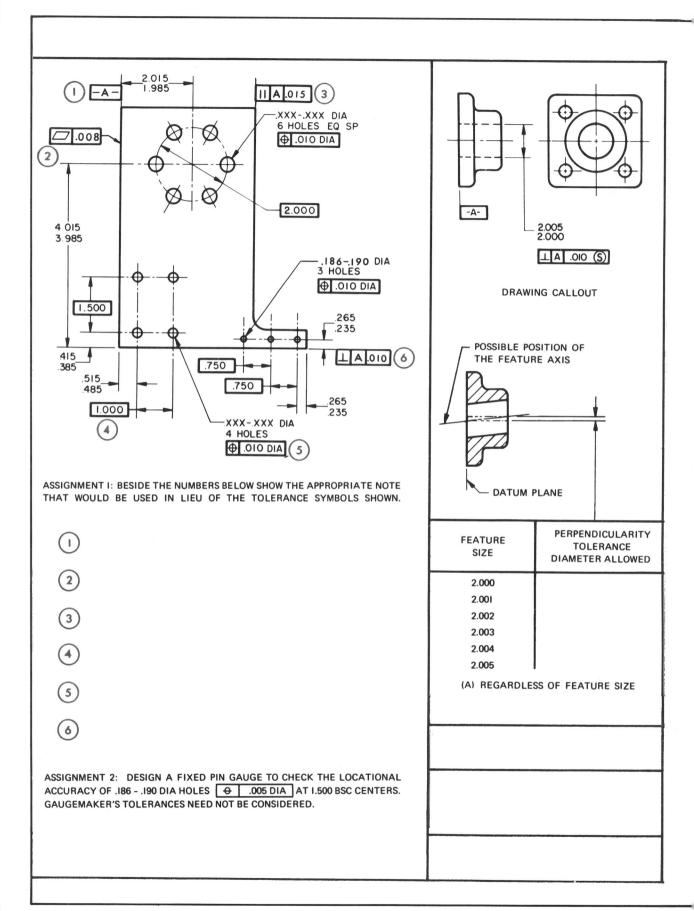

ASSIGNMENT I: BESIDE THE NUMBERS BELOW SHOW THE APPROPRIATE NOTE
THAT WOULD BE USED IN LIEU OF THE TOLERANCE SYMBOLS SHOWN.

1

2

3

4

5

6

ASSIGNMENT 2: DESIGN A FIXED PIN GAUGE TO CHECK THE LOCATIONAL
ACCURACY OF .186 – .190 DIA HOLES ⊕ .005 DIA AT 1.500 BSC CENTERS.
GAUGEMAKER'S TOLERANCES NEED NOT BE CONSIDERED.

DRAWING CALLOUT

POSSIBLE POSITION OF
THE FEATURE AXIS

DATUM PLANE

FEATURE SIZE	PERPENDICULARITY TOLERANCE DIAMETER ALLOWED
2.000	
2.001	
2.002	
2.003	
2.004	
2.005	

(A) REGARDLESS OF FEATURE SIZE

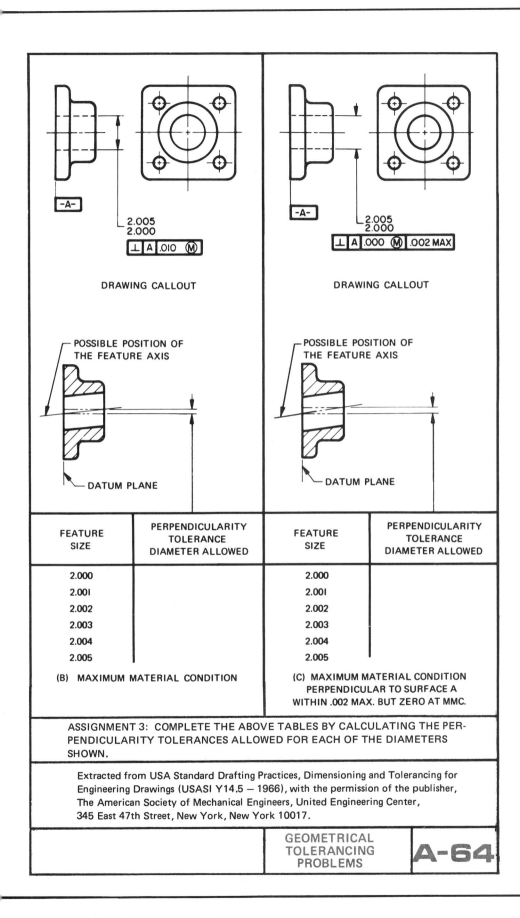

DRAWING CALLOUT

2.005
2.000

⊥ A .OIO Ⓜ

DRAWING CALLOUT

2.005
2.000

⊥ A .000 Ⓜ .002 MAX

POSSIBLE POSITION OF
THE FEATURE AXIS

DATUM PLANE

POSSIBLE POSITION OF
THE FEATURE AXIS

DATUM PLANE

FEATURE SIZE	PERPENDICULARITY TOLERANCE DIAMETER ALLOWED	FEATURE SIZE	PERPENDICULARITY TOLERANCE DIAMETER ALLOWED
2.000		2.000	
2.001		2.001	
2.002		2.002	
2.003		2.003	
2.004		2.004	
2.005		2.005	

(B) MAXIMUM MATERIAL CONDITION

(C) MAXIMUM MATERIAL CONDITION
PERPENDICULAR TO SURFACE A
WITHIN .002 MAX. BUT ZERO AT MMC.

ASSIGNMENT 3: COMPLETE THE ABOVE TABLES BY CALCULATING THE PER-
PENDICULARITY TOLERANCES ALLOWED FOR EACH OF THE DIAMETERS
SHOWN.

Extracted from USA Standard Drafting Practices, Dimensioning and Tolerancing for
Engineering Drawings (USASI Y14.5 – 1966), with the permission of the publisher,
The American Society of Mechanical Engineers, United Engineering Center,
345 East 47th Street, New York, New York 10017.

	GEOMETRICAL TOLERANCING PROBLEMS	A-64

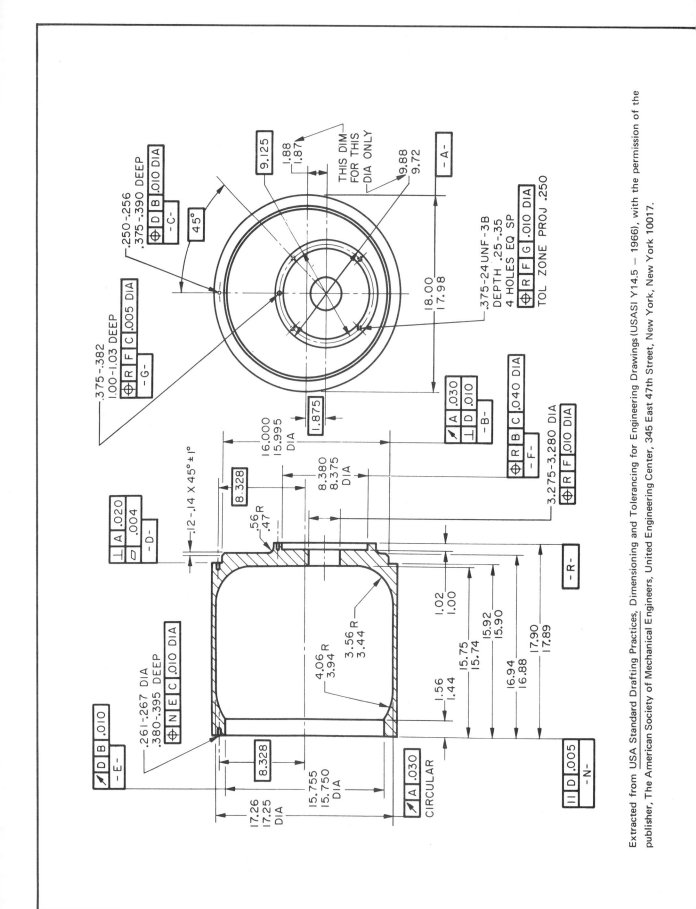

Extracted from USA Standard Drafting Practices, Dimensioning and Tolerancing for Engineering Drawings (USASI Y14.5 — 1966), with the permission of the publisher, The American Society of Mechanical Engineers, United Engineering Center, 345 East 47th Street, New York, New York 10017.

222

SKETCHING ASSIGNMENTS

1. Make a simple sketch of datum surface **D** showing the geometrical tolerances required.

2. Make a simple sketch of datum surface **N** showing the geometrical tolerances required.

QUESTIONS

1. How many datum surfaces or points are indicated?

2. How many true position dimensions are indicated?

3. How many datum surfaces are flat?

4. How many datum surfaces are circular?

5. How many dimensions show a true position tolerancing?

6. How many form tolerances are required?

7. How many different form tolerances are shown?

8. How many true position tolerances use datum **A** as a reference?

ANSWERS

1. _____

2. _____

3. _____

4. _____

5. _____

6. _____

7. _____

8. _____

MATERIAL	$\overline{C\,I}$
QUANTITY	104
SCALE	1/8
DRAWN	DATE

CASING	A-65

ROLLER-ELEMENT BEARINGS [1]

These bearings use some type of rolling element between the loaded members. Relative motion is accommodated by rotation of the elements. Bearing races which conform to the element shapes are normally used to house them. In addition, a cage or separator is often used to locate the elements within the bearings. These bearings are usually categorized by the form of the rolling element, and in some instances by the load type they carry. Roller-element bearings are generally classified as either ball or roller.

Ball Bearings. Ball bearings may be roughly divided into three categories: radial, angular contact, and thrust. Radial-contact ball bearings are designed for applications in which the load is primarily radial with only low-magnitude thrust loads. Angular-contact bearings are used where loads are combined radial and high thrust, and where precise shaft location is required. Thrust bearings handle loads which are primarily thrust.

Roller Bearings. Roller bearings have higher load capacities than ball bearings for a given envelope size. They are widely used in moderate-speed, heavy-duty applications. The four principle types of roller bearings are: cylindrical, needle, tapered, and spherical. Cylindrical roller bearings utilize cylinders with approximate length-diameter ratios ranging from 1:1 to 1:3 as rolling elements. Needle roller bearings utilize cylinders or needles of greater length-diameter ratios. Tapered and spherical roller bearings are capable of supporting combined radial and thrust loads

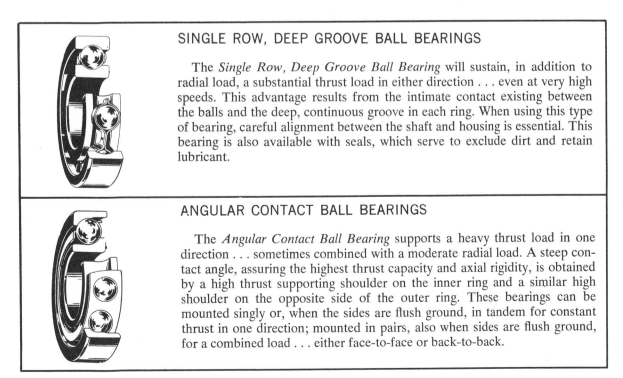

SINGLE ROW, DEEP GROOVE BALL BEARINGS

The *Single Row, Deep Groove Ball Bearing* will sustain, in addition to radial load, a substantial thrust load in either direction . . . even at very high speeds. This advantage results from the intimate contact existing between the balls and the deep, continuous groove in each ring. When using this type of bearing, careful alignment between the shaft and housing is essential. This bearing is also available with seals, which serve to exclude dirt and retain lubricant.

ANGULAR CONTACT BALL BEARINGS

The *Angular Contact Ball Bearing* supports a heavy thrust load in one direction . . . sometimes combined with a moderate radial load. A steep contact angle, assuring the highest thrust capacity and axial rigidity, is obtained by a high thrust supporting shoulder on the inner ring and a similar high shoulder on the opposite side of the outer ring. These bearings can be mounted singly or, when the sides are flush ground, in tandem for constant thrust in one direction; mounted in pairs, also when sides are flush ground, for a combined load . . . either face-to-face or back-to-back.

Fig. 41-1A Roller Element Bearings. (Courtesy Canadian SKF Co., Ltd.)

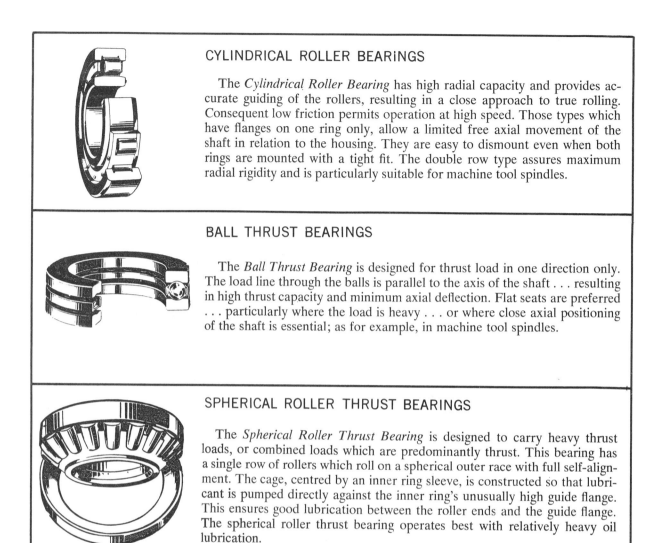

CYLINDRICAL ROLLER BEARINGS

The *Cylindrical Roller Bearing* has high radial capacity and provides accurate guiding of the rollers, resulting in a close approach to true rolling. Consequent low friction permits operation at high speed. Those types which have flanges on one ring only, allow a limited free axial movement of the shaft in relation to the housing. They are easy to dismount even when both rings are mounted with a tight fit. The double row type assures maximum radial rigidity and is particularly suitable for machine tool spindles.

BALL THRUST BEARINGS

The *Ball Thrust Bearing* is designed for thrust load in one direction only. The load line through the balls is parallel to the axis of the shaft . . . resulting in high thrust capacity and minimum axial deflection. Flat seats are preferred . . . particularly where the load is heavy . . . or where close axial positioning of the shaft is essential; as for example, in machine tool spindles.

SPHERICAL ROLLER THRUST BEARINGS

The *Spherical Roller Thrust Bearing* is designed to carry heavy thrust loads, or combined loads which are predominantly thrust. This bearing has a single row of rollers which roll on a spherical outer race with full self-alignment. The cage, centred by an inner ring sleeve, is constructed so that lubricant is pumped directly against the inner ring's unusually high guide flange. This ensures good lubrication between the roller ends and the guide flange. The spherical roller thrust bearing operates best with relatively heavy oil lubrication.

Fig. 41-1B   Roller Element Bearings, Continued.   (Courtesy Canadian SKF Co., Ltd.)

The rolling elements of tapered roller bearings are truncated cones. Spherical roller bearings are available with both barrel and hourglass roller shapes. The primary advantage of spherical roller bearings is their self-aligning capability.

RETAINING RINGS [2]

Retaining rings, or snap rings, are designed to provide a removable shoulder to locate, retain, or lock components accurately on shafts and in bores and housings. They are easily installed and removed, and since they are usually made of spring steel, retaining rings have a high shear strength and impact capacity. In addition to fastening and poisitioning, a number of rings are designed for taking up end play caused by accumulated tolerances or wear in the parts being retained.

O-RING SEALS

O-rings are used as an axial mechanical seal (a seal which forms a running seal between a moving shaft and a housing) or a static seal (no moving parts). The advantages of using an o-ring as a gasket-type seal over conventional

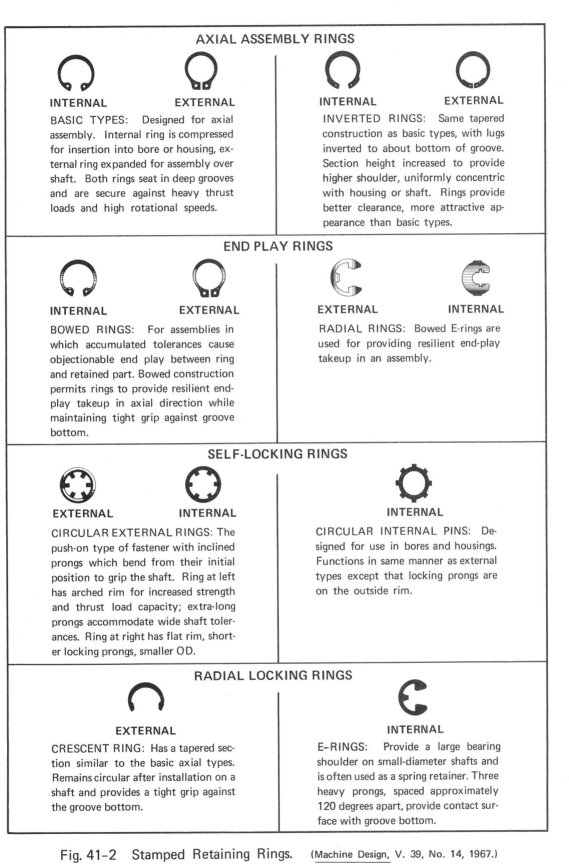

AXIAL ASSEMBLY RINGS

INTERNAL   EXTERNAL

BASIC TYPES: Designed for axial assembly. Internal ring is compressed for insertion into bore or housing, external ring expanded for assembly over shaft. Both rings seat in deep grooves and are secure against heavy thrust loads and high rotational speeds.

INTERNAL   EXTERNAL

INVERTED RINGS: Same tapered construction as basic types, with lugs inverted to about bottom of groove. Section height increased to provide higher shoulder, uniformly concentric with housing or shaft. Rings provide better clearance, more attractive appearance than basic types.

END PLAY RINGS

INTERNAL   EXTERNAL

BOWED RINGS: For assemblies in which accumulated tolerances cause objectionable end play between ring and retained part. Bowed construction permits rings to provide resilient end-play takeup in axial direction while maintaining tight grip against groove bottom.

EXTERNAL   INTERNAL

RADIAL RINGS: Bowed E-rings are used for providing resilient end-play takeup in an assembly.

SELF-LOCKING RINGS

EXTERNAL   INTERNAL

CIRCULAR EXTERNAL RINGS: The push-on type of fastener with inclined prongs which bend from their initial position to grip the shaft. Ring at left has arched rim for increased strength and thrust load capacity; extra-long prongs accommodate wide shaft tolerances. Ring at right has flat rim, shorter locking prongs, smaller OD.

INTERNAL

CIRCULAR INTERNAL PINS: Designed for use in bores and housings. Functions in same manner as external types except that locking prongs are on the outside rim.

RADIAL LOCKING RINGS

EXTERNAL

CRESCENT RING: Has a tapered section similar to the basic axial types. Remains circular after installation on a shaft and provides a tight grip against the groove bottom.

INTERNAL

E-RINGS: Provide a large bearing shoulder on small-diameter shafts and is often used as a spring retainer. Three heavy prongs, spaced approximately 120 degrees apart, provide contact surface with groove bottom.

Fig. 41-2 Stamped Retaining Rings. (Machine Design, V. 39, No. 14, 1967.)

gaskets is that the nuts need not be tightened uniformly and sealing dopes are not required. A rectangular groove is the most common type of groove used for o-rings.

CLUTCHES

Clutches are used to start and stop a machine or rotating element without starting or stopping the prime mover. They are also used for automatic disconnection, quick starts and stops and to permit shaft rotation in one direction only such as the *over-running* clutch shown in figure 41-4. A full complement of

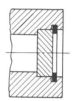

EXTERNAL

(A) AXIAL AND
RADIAL ASSEMBLY

INTERNAL

(B) AXIAL ASSEMBLY

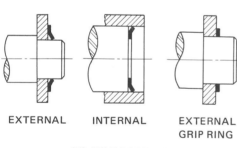

EXTERNAL   INTERNAL   EXTERNAL
GRIP RING

(C) SELF-LOCKING

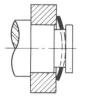
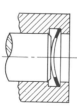

EXTERNAL   INTERNAL

(D) END-PLAY TAKEUP

Fig. 41-3   Retaining Ring Application.
(Courtesy Waldes Kohinoor, Inc. (c) 1958, 1965,
Reprinted With Permission.)

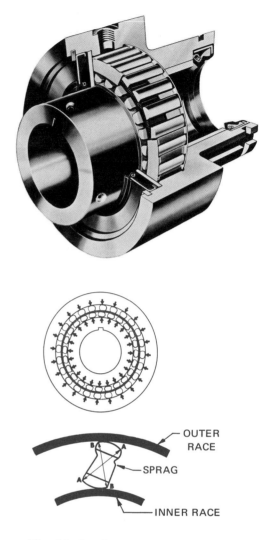

Fig. 41-4   Over-running Clutch.
(Courtesy Formsprag Co.)

sprags between concentric inner and outer races transmits power from one race to the other by wedging action of the sprags when either race is rotated in the driving direction. Rotation in the opposite direction frees the sprags and the clutch is disengaged or *over-runs.* This type of clutch is used in the power drive, drawing A-66.

BELT DRIVES [3]

A belt drive consists of an endless flexible belt that connects two wheels, or pulleys. Belt drives depend on friction between belt and pulley surfaces for transmission of power.

In a V-belt drive, the belt has a trapezoidal cross section, and runs in V-shaped grooves on the pulleys. These belts are made up of cords or cables, impregnated and covered with rubber or other organic compound. The covering is formed to produce the required cross section. V-belts are usually manufactured as endless belts, although open-end and link types are available.

In the case of V-belts, the friction for the transmission of the driving force is increased by the wedging of the belt into the grooves on the pulley.

Fig. 41-6   V-Belt Sheave and Bushing.
(Courtesy T.B. Woods' Sons Co.)

V-Belt Sizes

To facilitate interchangeability and to insure uniformity, V-belt manufacturers have developed industrial standards for the various types of V-belts. Industrial V-belts are made in two types: heavy duty (conventional and narrow) and light duty. Conventional belts are available in A, B, C, D, and E sections. Narrow belts are made in 3V, 5V, and 8V sections. Light-duty or fractional-horsepower belts come in 3L, 4L, and 5L sections.

Sheaves and Bushings

Sheaves (the grooved wheels of pulleys) are sometimes equipped with tapered bushings for ease of installation and removal. They have extreme holding power, providing the equivalent of a shrink fit. The sheave and bushing used in the power drive, drawing A-66, have a six-hole drilling arrangement in both the bushing and sheave making it possible to insert the cap screw from either side. This is especially advantageous for applications where space is at a premium.

REFERENCES AND SOURCE MATERIAL

1. A.O. Dehayt, "Basic Bearing Types," *Machine Design* 40, No. 14 (1968).

2. *Machine Design* 39, No. 14 (1967).

3. American Sprocket Chain Manufacturers' Association.

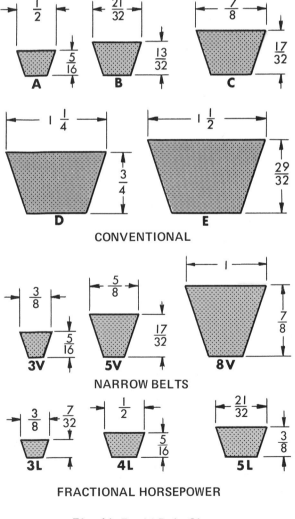

CONVENTIONAL

NARROW BELTS

FRACTIONAL HORSEPOWER

Fig. 41-5   V-Belt Sizes.

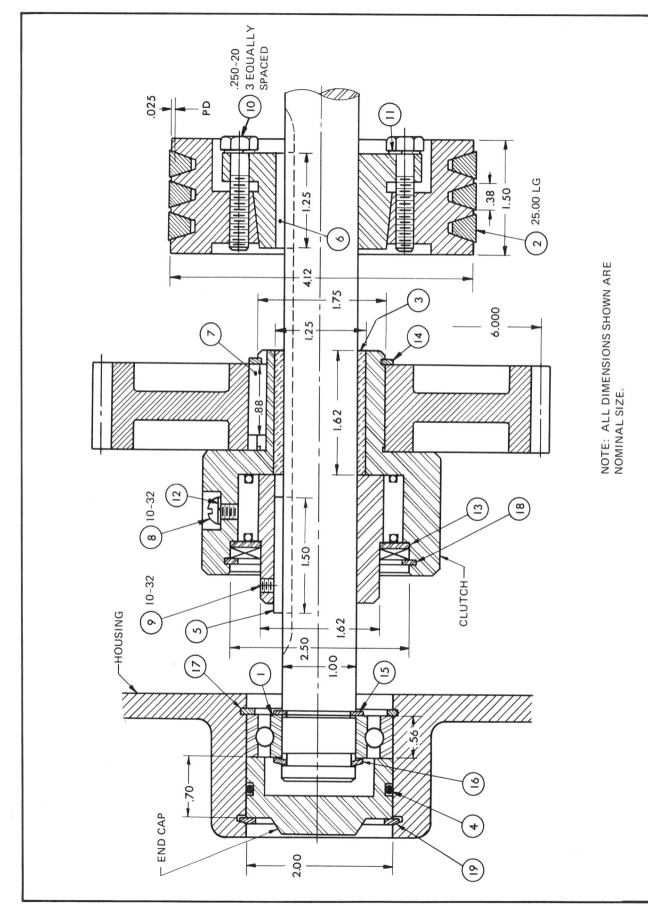

NOTE: ALL DIMENSIONS SHOWN ARE
NOMINAL SIZE.

MAKE A DETAIL DRAWING OF THE
END CAP –.10 SQUARES

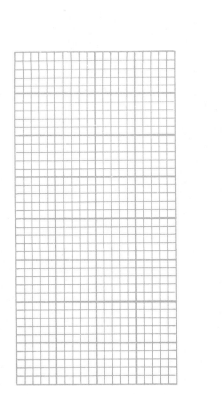

SKETCHING ASSIGNMENT

In the sketching area provided, make a one-view detail drawing of the end cap, using dimensions taken from o-ring catalogs for the groove. Use your judgment for dimensions not shown.

BILL OF MATERIAL

Prepare a bill of material for parts I to 19. Refer to manufacturers' catalogs and tables found in drafting manuals.

QUESTIONS

1. How many cap screws fasten the sheave to the bushing?

2. List five parts or methods that are used to lock or join parts together on this assembly.

3. How many V-belts are used?

4. How many keys are there?

5. How many retaining rings are used?

6. What type of bearing is part ③ ?

7. What type of bearing is part ① ?

8. What prevents the oil from leaking out between the housing and the end cap?

9. Can the gear be driven in both directions?

10. If the diametral pitch on the gear is 8, what is the number of teeth?

11. What is the pitch diameter of the sheave?

12. What size of V-belt is required?

ANSWERS

1 _____

2 _____

3 _____

4 _____

5 _____

6 _____

7 _____

8 _____

9 _____

10 _____

11 _____

12 _____

SCALE	1/1	
DRAWN		DATE
POWER DRIVE		**A-66**

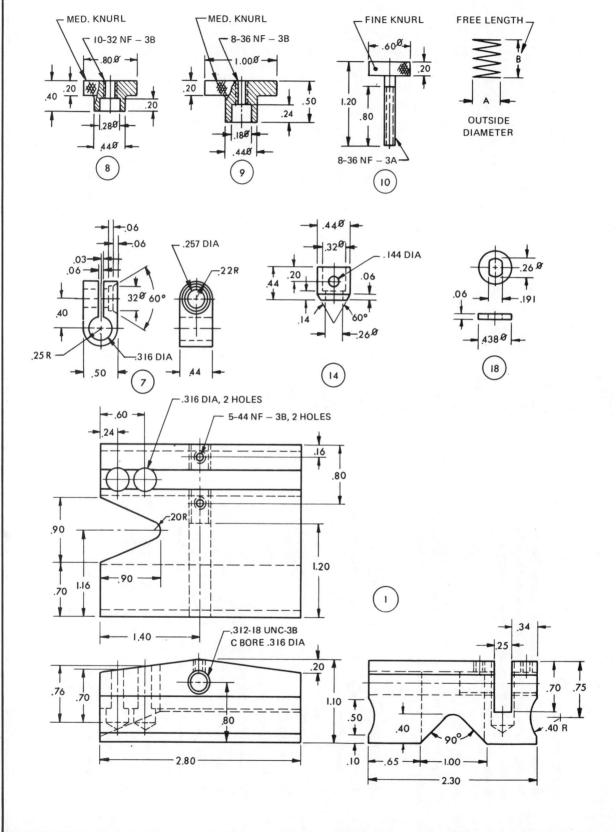

MED. KNURL

10-32 NF – 3B

.80∅

.40   .20

.20

.28∅

.44∅

⑧

MED. KNURL

8-36 NF – 3B

1.00∅

.20

.50

.24

.180∅

.44∅

⑨

FINE KNURL

.60∅

.20

1.20

.80

8-36 NF – 3A

⑩

FREE LENGTH

B

A

OUTSIDE
DIAMETER

.06

.06

.03

.06

.257 DIA

.22R

32∅ 60°

.40

.25 R

.50

.316 DIA

.44

⑦

.44∅

.32∅

.20

.44

.14

.144 DIA

.06

60°

.26∅

⑭

.26∅

.06

.191

.438∅

⑱

.60

.24

.316 DIA, 2 HOLES

5-44 NF – 3B, 2 HOLES

.16

.80

.90

.20R

.70

1.16

.90

1.40

1.20

①

.312-18 UNC-3B
C BORE .316 DIA

.76

.70

.20

.80

1.10

.50

.10

2.80

.34

.25

.70   .75

.40

90°

.40 R

.65   1.00

2.30

232

PT.	A	B	PITCH
15	.30	.70	.10
16	.25	32	.08
17	.25	.40	.08

SPRINGS

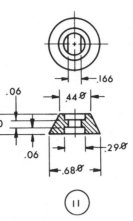

.144 DIA

10-32 NF – 3A

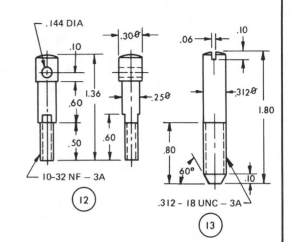

⑫

.312 – 18 UNC – 3A

⑬

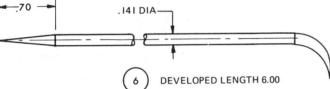

⑥ DEVELOPED LENGTH 6.00

.316 DIA   .144 DIA

60°

④

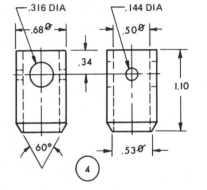

8-36 NF – 3B x .50 LG

.312∅

7.50

⑤

.316 DIA   .144 DIA

.310

8-36 NF – 3A

③

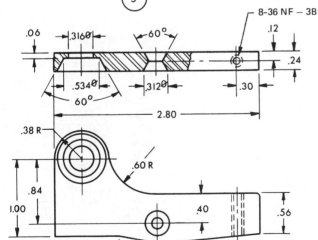

8-36 NF – 3B

.38 R

.60 R

60°

2.80

1.00   .84

.40

.56

.10   .80   .60

②

NOTE: FOR SURFACE GAUGE ASSEMBLY SEE DRAWING A–68

SCALE	1/1	
DRAWN		DATE

SURFACE GAUGE
DETAILS

A-67

233

ASSIGNMENT

1. Each individual part of the surface gauge is designated by a part number on the working drawing and in the bill of materials. Identify each of these parts on the assembly drawing.

2. Complete the bill of materials.

QUESTIONS

1. What part maintains the frictional contact between part ④ and part ⑤ when thumbnut ⑨ is loosened?

2. What part is actuated so that scriber ⑥ may be adjusted on spindle ⑤ ?

3. What part keeps spindle ⑤ from being pulled through parts ③ and ④ ?

4. What part actuates micrometer adjustment of rocker ② to raise or lower scriber ⑥ ?

5. What part or parts keep rocker ② in tension when adjustment is being made with thumbscrew ⑩ ?

6. What part or parts serve as pivots for rocker ② ?

7. What is the length of the setscrews at Ⓐ ?

8. What is the outside diameter of the spring at Ⓑ ?

9. What is the diameter of hole Ⓒ ?

10. Determine radius Ⓛ .

11. What is the size of thread on the thumbscrew Ⓗ ?

12. What is the developed length of part ⑥ ?

13. What is the diameter of the spring wire for part Ⓙ ?

14. What is the length of part Ⓚ ?

15. Calculate distances Ⓓ Ⓔ Ⓕ Ⓖ .

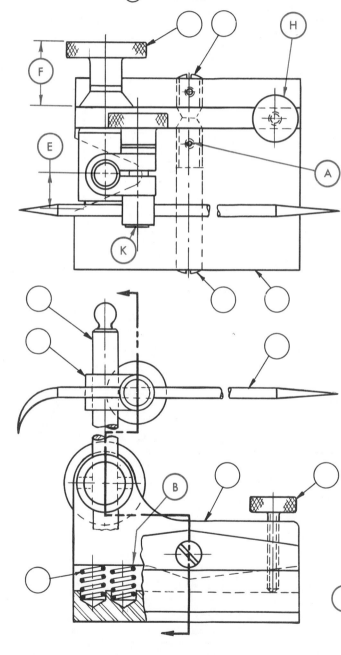

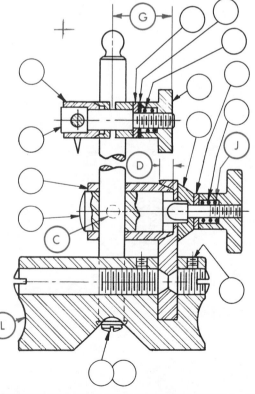

NOTE: FOR SURFACE GAUGE DETAILS SEE DRAWING A—67

ANSWERS

1 _____

2 _____

3 _____

4 _____

5 _____

6 _____

7 _____

8 _____

9 _____

10 _____

11 _____

12 _____

13 _____

14 _____

15 (D) _____

(E) _____

(F) _____

(G) _____

QTY.	ITEM	MATL.	DESCRIPTION	PT. NO.
				30
				29
				28
				27
				26
				25
				24
1	MACHINE SCREW	ST	PANHEAD 8–36 UNF x 38 LG.	23
2	SET SCREW	ST	SLOTTED HEADLESS CUP POINT 5-44 NF x .18 LG.	22
1	SET SCREW	ST	SLOTTED HEADLESS CUP POINT .312 UNC x .50 LG.	21
1	WASHER	ST	.219 ID x .438 OD x .032	20
2	WASHER	ST	.188 ID x .438 OD x .049	19
1	WASHER	ST		18
1	SPRING	ST	12 A.S.W. MUSIC WIRE	17
1	SPRING	ST	12 A.S.W. MUSIC WIRE	16
2	SPRING	ST	23 A.S.W. MUSIC WIRE	15
1	SPACER	ST		14
1	STUD	ST		13
1	LOCK SCREW	ST		12
1	SPACER	ST		11
1	THUMB SCREW	ST		10
1	THUMB NUT	ST		9
1	THUMB NUT	ST		8
1	SWIVEL	ST		7
1	SCRIBER	ST		6
1	SPINDLE	ST		5
1	SPACER	ST		4
1	SCREW ARM	ST		3
1	ROCKER	ST		2
1	BASE	ST		1
QTY.	ITEM	MATL.	DESCRIPTION	PT. NO.

SCALE	1–1	
DRAWN		DATE

SURFACE GAUGE ASSEMBLY **A-68**

235

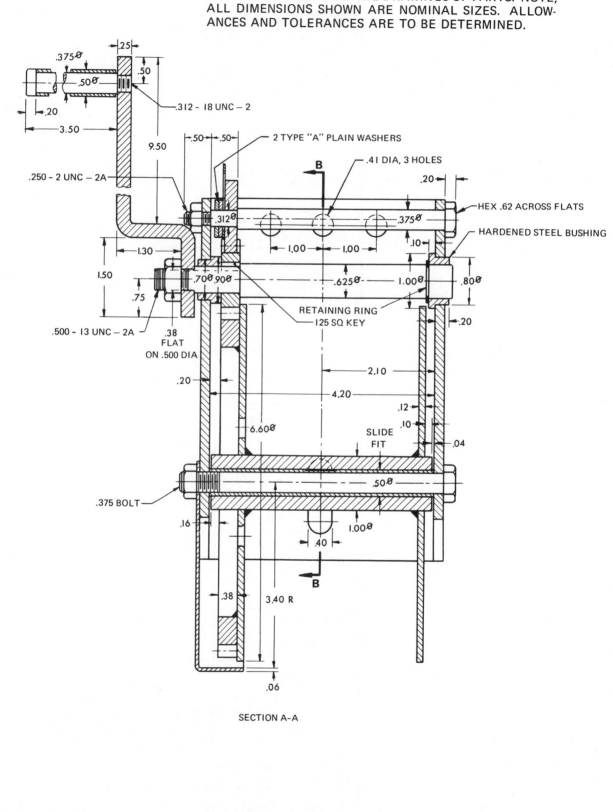

ASSIGNMENT: MAKE DETAIL DRAWINGS OF PARTS. NOTE,
ALL DIMENSIONS SHOWN ARE NOMINAL SIZES. ALLOW-
ANCES AND TOLERANCES ARE TO BE DETERMINED.

.375∅

.50∅

.25

.375∅

.50

.20

3.50

.312 – 18 UNC – 2

9.50

2 TYPE "A" PLAIN WASHERS

.50 .50

.41 DIA, 3 HOLES

.20

.250 – 2 UNC – 2A

.312∅

.375∅

HEX .62 ACROSS FLATS

HARDENED STEEL BUSHING

1.30

1.00 1.00

.10

1.50

.700 .900

.625∅

1.00∅

.80∅

.75

RETAINING RING
125 SQ KEY

.20

.500 – 13 UNC – 2A

.38
FLAT
ON .500 DIA

.20

2.10

4.20

.12

.10

.04

6.60∅

SLIDE
FIT

.50∅

.375 BOLT

.16

1.00∅

.40

.38

3.40 R

B

B

.06

SECTION A-A

COURTESY FULTON COMPANY, MILWAUKEE, WISCONSIN, and J.C. ADAMS COMPANY LTD., REXDALE, ONTARIO, CANADA

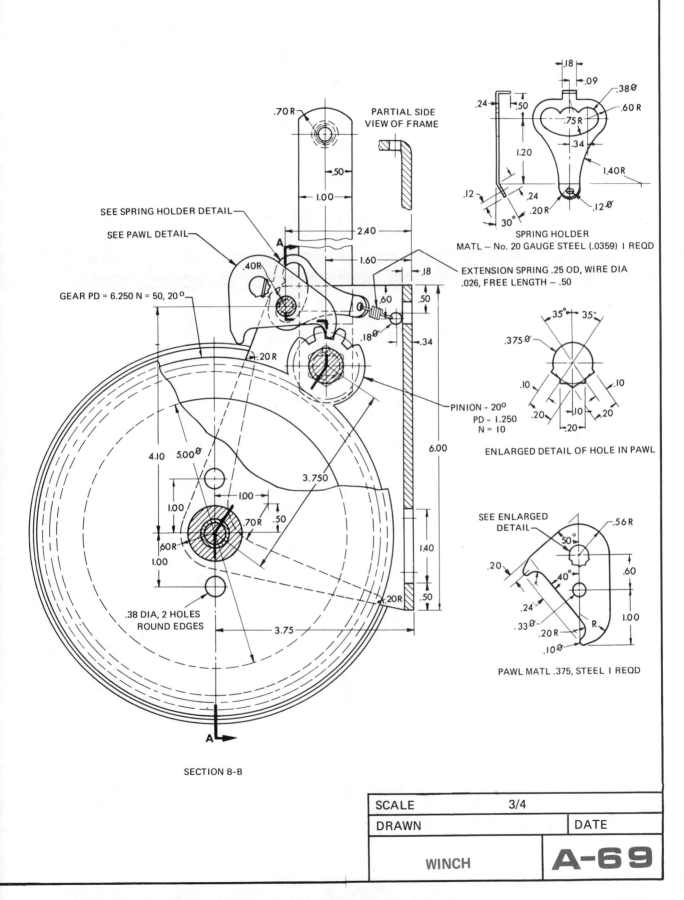

.70R

PARTIAL SIDE
VIEW OF FRAME

.50

1.00

2.40

SEE SPRING HOLDER DETAIL

SEE PAWL DETAIL

A

1.60

.40R

.18

GEAR PD = 6.250 N = 50, 20°

.60

.50

.18⌀

.34

.20R

PINION - 20°
PD - 1.250
N = 10

6.00

4.10

5.00⌀

3.750

1.00

1.00

.70R

.50

.60R

1.40

1.00

.20R

.50

.38 DIA, 2 HOLES
ROUND EDGES

3.75

A

SECTION B-B

.18

.09

.38⌀

.60R

.75R

.24

.50

1.20

.34

1.40R

.12

.24

.20R

.12⌀

30°

SPRING HOLDER
MATL — No. 20 GAUGE STEEL (.0359) I REQD

EXTENSION SPRING .25 OD, WIRE DIA
.026, FREE LENGTH — .50

35° 35°

.375⌀

.10

.10

.20

.10

.20

.20

ENLARGED DETAIL OF HOLE IN PAWL

SEE ENLARGED
DETAIL

.56R

50°

.20

40°

.60

.24

R

1.00

.33⌀

.20R

.10⌀

PAWL MATL .375, STEEL I REQD

SCALE	3/4	
DRAWN		DATE
WINCH		A-69

237

FRACTION	DECIMALS		FRACTION	DECIMALS	
	TWO PLACE	THREE PLACE		TWO PLACE	THREE PLACE
1/64	.02	.016	33/64	.52	.516
1/32	.03	.031	17/32	.53	.531
3/64	.05	.047	35/64	.55	.547
1/16	.06	.062	9/16	.56	.562
5/64	.08	.078	37/64	.58	.578
3/32	.09	.094	19/32	.59	.594
7/64	.11	.109	39/64	.61	.609
1/8	.12	.125	5/8	.62	.625
9/64	.14	.141	41/64	.64	.641
5/32	.16	.156	21/32	.66	.656
11/64	.17	.172	43/64	.67	.672
3/16	.19	.188	11/16	.69	.688
13/64	.20	.203	45/64	.70	.703
7/32	.22	.219	23/32	.72	.719
15/64	.23	.234	47/64	.73	.734
1/4	.25	.250	3/4	.75	.750
17/64	.27	.266	49/64	.77	.766
9/32	.28	.281	25/32	.78	.781
19/64	.30	.297	51/64	.80	.797
5/16	.31	.312	13/16	.81	.812
21/64	.33	.328	53/64	.83	.828
11/32	.34	.344	27/32	.84	.844
23/64	.36	.359	55/64	.86	.859
3/8	.38	.375	7/8	.88	.875
25/64	.39	.391	57/64	.89	.891
13/32	.41	.406	29/32	.91	.906
27/64	.42	.422	59/64	.92	.922
7/16	.44	.438	15/16	.94	.938
29/64	.45	.453	61/64	.95	.953
15/32	.47	.469	31/32	.97	.969
31/64	.48	.484	63/64	.98	.984
1/2	.50	.500	1	1.00	1.000

Table 1   Conversion Chart For Converting Fractional Dimensions to Decimal Dimensions.

Size	COARSE THREAD SERIES UNC & NC		FINE THREAD SERIES UNF & NF		EXTRA FINE THREAD SERIES UNEF & NEF		8 – PITCH THREAD SERIES 8 N		12 – PITCH THREAD SERIES 12 N		16 – PITCH THREAD SERIES 16 N	
	Threads Per Inch	Tap Drill	Threads Per Inch	Tap Drill	Threads Per Inch	Tap Drill	Threads Per Inch	Tap Drill	Threads Per Inch	Tap Drill	Threads Per Inch	Tap Drill
.060			80	3/64								
1 .073	64	No. 53	72	No. 53								
2 .086	56	No. 50	64	No. 50								
3 .099	48	No. 47	56	No. 45								
4 .112	40	No. 43	48	No. 42								
5 .125	40	No. 38	44	No. 37								
6 .138	32	No. 36	40	No. 33								
8 .164	32	No. 29	36	No. 29								
10 .190	24	No. 25	32	No. 21								
12 .216	24	No. 16	28	No. 14	32	No. 13						
.250	20	No. 7	28	No. 3	32	7/32						
.312	18	F	24	I	32	9/32						
.375	16	5/16	24	Q	32	11/32						
.438	14	U	20	25/64	28	13/32						
.500	13	27/64										
.500	12		20	29/64	28	15/32			12	27/64		
.562	12	31/64	18	33/64	24	33/64			12	31/64		
.625	11	17/32	18	37/64	24	37/64			12	35/64		
.750	10	21/32	16	11/16	20	45/64			12	43/64	16	11/16
.875	9	49/64	14	13/16	20	53/64			12	51/64	16	13/16
1.000	8	7/8	12	59/64	20	61/64	8	7/8	12	59/64	16	15/16
1.125	7	63/64	12	1 3/64	18	1 5/64	8	1	12	1 3/64	16	1 1/16
1.250	7	1 7/64	12	1 11/64	18	1 3/16	8	1 1/8	12	1 11/64	16	1 3/16
1.375	6	1 7/32	12	1 19/64	18	1 5/16	8	1 1/4	12	1 19/64	16	1 5/16
1.500	6	1 11/32	12	1 27/64	18	1 7/16	8	1 3/8	12	1 27/64	16	1 7/16
1.750	5	1 9/16			16	1 11/16	8	1 5/8	12	1 43/64	16	1 11/16
2.000	4 1/2	1 25/32			16	1 15/16	8	1 7/8	12	1 51/64	16	1 15/16
2.250	4 1/2	2 1/32					8	2 1/8	12	2 11/64	16	2 3/16
2.500	4	2 1/4					8	2 3/8	12	2 27/64	16	2 7/16
2.750	4	2 1/2					8	2 5/8	12	2 43/64	16	2 11/16
3.000	4	2 3/4					8	2 7/8	12	2 59/64	16	2 15/16
3.250	4	3					8	3 1/8	12	3 11/64	16	3 3/16
3.500	4	3 1/4					8	3 3/8	12	3 27/64	16	3 7/16
3.750	4	3 1/2					8	3 5/8	12	3 43/64	16	3 11/16
4.000	4	3 3/4					8	3 7/8	12	3 59/64	16	3 15/16
4.250							8	4 1/8	12	4 11/64	16	4 3/16
4.500							8	4 3/8	12	4 27/64	16	4 7/16
4.750							8	4 5/8	12	4 43/64	16	4 11/16
5.000							8	4 7/8	12	4 59/64	16	4 15/16
5.250							8	5 1/8	12	5 11/64	16	5 3/16
5.500							8	5 3/8	12	5 27/64	16	5 7/16
5.750							8	5 5/8	12	5 43/64	16	5 11/16
6.000							8	5 7/8	12	5 59/64	16	5 15/16

Color shows unified threads CSA B1.1 – 1949 Addendum No. 1 – 1951. Information identical to that found in ASA B1.1 – 1960.

Table 2 Unified and American Threads.(CSA – B1.1 – 1949)

Table 3 Conversion of Fractions of an Inch to Millimeters (1 Inch = 25.4 Millimeters).

IN.	0	1/16	1/8	3/16	1/4	5/16	3/8	7/16	1/2	9/16	5/8	11/16	3/4	13/16	7/8	15/16
0	.0	1.6	3.2	4.8	6.4	7.9	9.5	11.1	12.7	14.3	15.9	17.5	19.1	20.6	22.2	23.8
1	25.4	27.0	28.6	30.2	31.8	33.3	34.9	36.5	38.1	39.7	41.3	42.9	44.5	46.0	47.6	49.2
2	50.8	52.4	54.0	55.6	57.2	58.7	60.3	61.9	63.5	65.1	66.7	68.3	69.9	71.4	73.0	74.6
3	76.2	77.8	79.4	81.0	82.6	84.1	85.7	87.3	88.9	90.5	92.1	93.7	95.3	96.8	98.4	100.0
4	101.6	103.2	104.8	106.4	108.0	109.5	111.1	112.7	114.3	115.9	117.5	119.1	120.7	122.2	123.8	125.4
5	127.0	128.6	130.2	131.8	133.4	134.9	136.5	138.1	139.7	141.3	142.9	144.5	146.1	147.6	149.2	150.8
6	152.4	154.0	155.6	157.2	158.8	160.3	161.9	163.5	165.1	166.7	168.3	169.9	171.5	173.0	174.6	176.2
7	177.8	179.4	181.0	182.6	184.2	185.7	187.3	188.9	190.5	192.1	193.7	195.3	196.9	198.4	200.0	201.6
8	203.2	204.8	206.4	208.0	209.6	211.1	212.7	214.3	215.9	217.5	219.1	220.7	222.3	223.8	225.4	227.0
9	228.6	230.2	231.8	233.4	235.0	236.5	238.1	239.7	241.3	242.9	244.5	246.1	247.7	249.2	250.8	252.4
10	254.0	255.6	257.2	258.8	260.4	261.9	263.5	265.1	266.7	268.3	269.9	271.5	273.1	274.6	276.2	277.8
11	279.4	281.0	282.6	284.2	285.8	287.3	288.9	290.5	292.1	293.7	295.3	296.9	298.5	300.0	301.6	303.2
12	304.8	306.4	308.0	309.6	311.2	312.7	314.3	315.9	317.5	319.1	320.7	322.3	323.9	325.4	327.0	328.6
13	330.2	331.8	333.4	335.0	336.6	338.1	339.7	341.3	342.9	344.5	346.1	347.7	349.3	350.8	352.4	354.0
14	355.6	357.2	358.8	360.4	362.0	363.5	365.1	366.7	368.3	369.9	371.5	373.1	374.7	376.2	377.8	379.4

Table 4 Conversion of Decimals of an Inch to Millimeters (.10 Inch = 2.54 Millimeters). Courtesy (Crane Canada Limited)

IN.	.00	.01	.02	.03	.04	.05	.06	.07	.08	.09	IN.
.00	.00	.25	.51	.76	1.02	1.27	1.52	1.78	2.03	2.29	.00
.10	2.54	2.79	3.05	3.30	3.56	3.81	4.06	4.32	4.57	4.83	.10
.20	5.08	5.33	5.59	5.84	6.10	6.35	6.60	6.86	7.11	7.37	.20
.30	7.62	7.87	8.13	8.38	8.64	8.89	9.14	9.40	9.65	9.91	.30
.40	10.16	10.41	10.67	10.92	11.18	11.43	11.68	11.94	12.19	12.45	.40
.50	12.70	12.95	13.21	13.46	13.72	13.97	14.22	14.48	14.73	14.99	.50
.60	15.24	15.49	15.75	16.00	16.26	16.51	16.76	17.02	17.27	17.53	.60
.70	17.78	18.03	18.29	18.54	18.80	19.05	19.30	19.56	19.81	20.07	.70
.80	20.32	20.57	20.83	21.08	21.34	21.59	21.84	22.10	22.35	22.61	.80
.90	22.86	23.11	23.37	23.62	23.88	24.13	24.38	24.64	24.89	25.15	.90

MM.	0	1	2	3	4	5	6	7	8	9	MM.
0	.00	.039	.079	.118	.157	.197	.236	.276	.315	.354	0
10	.39	.43	.47	.51	.55	.59	.63	.67	.71	.75	10
20	.79	.83	.87	.91	.94	.98	1.02	1.06	1.10	1.14	20
30	1.18	1.22	1.26	1.30	1.34	1.38	1.42	1.46	1.50	1.54	30
40	1.57	1.61	1.65	1.69	1.73	1.77	1.81	1.85	1.89	1.93	40
50	1.97	2.01	2.05	2.09	2.13	2.17	2.20	2.24	2.28	2.32	50
60	2.36	2.40	2.44	2.48	2.52	2.56	2.60	2.64	2.68	2.72	60
70	2.76	2.80	2.83	2.87	2.91	2.95	2.99	3.03	3.07	3.11	70
80	3.15	3.19	3.23	3.27	3.31	3.35	3.39	3.43	3.46	3.50	80
90	3.54	3.58	3.62	3.66	3.70	3.74	3.78	3.82	3.86	3.90	90
100	3.94	3.98	4.02	4.06	4.09	4.13	4.17	4.21	4.25	4.29	100
110	4.33	4.37	4.41	4.45	4.49	4.53	4.57	4.61	4.65	4.69	110
120	4.72	4.76	4.80	4.84	4.88	4.92	4.96	5.00	5.04	5.08	120
130	5.12	5.16	5.20	5.24	5.28	5.31	5.35	5.39	5.43	5.47	130
140	5.51	5.55	5.59	5.63	5.67	5.71	5.75	5.79	5.83	5.87	140
150	5.91	5.94	5.98	6.02	6.06	6.10	6.14	6.18	6.22	6.26	150
160	6.30	6.34	6.38	6.42	6.46	6.50	6.54	6.57	6.61	6.65	160
170	6.69	6.73	6.77	6.81	6.85	6.89	6.93	6.97	7.01	7.05	170
180	7.09	7.13	7.17	7.20	7.24	7.28	7.32	7.36	7.40	7.44	180
190	7.48	7.52	7.56	7.60	7.64	7.68	7.72	7.76	7.80	7.83	190
200	7.87	7.91	7.95	7.99	8.03	8.07	8.11	8.15	8.19	8.23	200
210	8.27	8.31	8.35	8.39	8.43	8.46	8.50	8.54	8.58	8.62	210
220	8.66	8.70	8.74	8.78	8.82	8.86	8.90	8.94	8.98	9.02	220
230	9.06	9.09	9.13	9.17	9.21	9.25	9.29	9.33	9.37	9.41	230
240	9.45	9.49	9.53	9.57	9.61	9.65	9.69	9.72	9.76	9.80	240

Table 5 Conversion of Millimeters to Inches (1 Millimeter = .03937 Inch).

DECIMAL EQUIVALENTS OF NUMBER SIZE DRILLS				DECIMAL EQUIVALENTS OF LETTER SIZE DRILLS	
NO.	DECIMAL EQUIVALENT	NO.	DECIMAL EQUIVALENT	LETTER	DECIMAL EQUIVALENT
1	.2280	31	.1200	A	.234
2	.2210	32	.1160	B	.238
3	.2130	33	.1130	C	.242
4	.2090	34	.1110	D	.246
5	.2055	35	.1100	E	.250
6	.2040	36	.1065	F	.257
7	.2010	37	.1040	G	.261
8	.1990	38	.1015	H	.266
9	.1960	39	.0995	I	.272
10	.1935	40	.0980	J	.277
11	.1910	41	.0960	K	.281
12	.1890	42	.0935	L	.290
13	.1850	43	.0890	M	.295
14	.1820	44	.0860	N	.302
15	.1800	45	.0820	O	.316
16	.1770	46	.0810	P	.323
17	.1730	47	.0785	Q	.332
18	.1695	48	.0760	R	.339
19	.1660	49	.0730	S	.348
20	.1610	50	.0700	T	.358
21	.1590	51	.0670	U	.368
22	.1570	52	.0635	V	.377
23	.1540	53	.0595	W	.386
24	.1520	54	.0550	X	.397
25	.1495	55	.0520	Y	.404
26	.1470	56	.0465	Z	.413
27	.1440	57	.0430		
28	.1405	58	.0420		
29	.1360	59	.0410		
30	.1285	60	.0400		

Table 6   Number and Letter–Size Drills.

ANSI PUBLICATIONS

ANSI Y14.1	Size and Format
ANSI Y14.2-1957	Line Conventions, Sectioning and Lettering
ANSI Y14.3-1957	Projections
ANSI Y14.4-1957	Pictorial Drawing
ANSI Y14.5-1966	Dimensioning and Tolerancing for Engineering Drawings
ANSI Y14.6-1957	Screw Threads
ANSI Y14.7	Gears, Splines and Serrations
ANSI Y14.7.1	Gear Drawing Standards — Part 1, For Spur, Helical, Double Helical and Rack
ANSI Y14.9	Forgings
ANSI Y14.10	Metal Stampings
ANSI Y14.11	Plastics
ANSI Y14.14	Mechanical Assemblies
ANSI Y14.15	Electrical and Electronics Diagrams
ANSI Y14.15A	Interconnection Diagrams
ANSI Y14.17	Fluid Power Diagrams
ANSI Y32.2	Graphic Symbols for Electrical and Electronics Diagrams
ANSI Y32.9	Graphic Electrical Wiring Symbols for Architectural and Electrical Layout Drawings
ANSI B1.1	Unified Screw Threads
ANSI B18.2.1	Square and Hex Bolts and Screws
ANSI B18.2.2	Square and Hex Nuts
ANSI B18.3	Socket Cap, Shoulder and Set Screws
ANSI B18.6.2	Slotted Head Cap Screws, Square Head Set Screws, Slotted Headless Set Screws
ANSI B18.6.3	Machine Screws and Machine Screw Nuts
ANSI B17.2	Woodruff Key and Keyslot Dimensions
ANSI B17.1	Keys and Keyseats
ANSI B18.21.1	Lock Washers
ANSI B27.2	Plain Washers
ANSI B46.1	Surface Texture

The Above Standards May Be Purchased From:

ANSI   American National Standards Institute, Inc.
1430 Broadway
New York, New York 10018

The American Society of Mechanical Engineers
United Engineering Center
345 East 47th Street
New York, New York 10017

Table 7   ANSI Publications